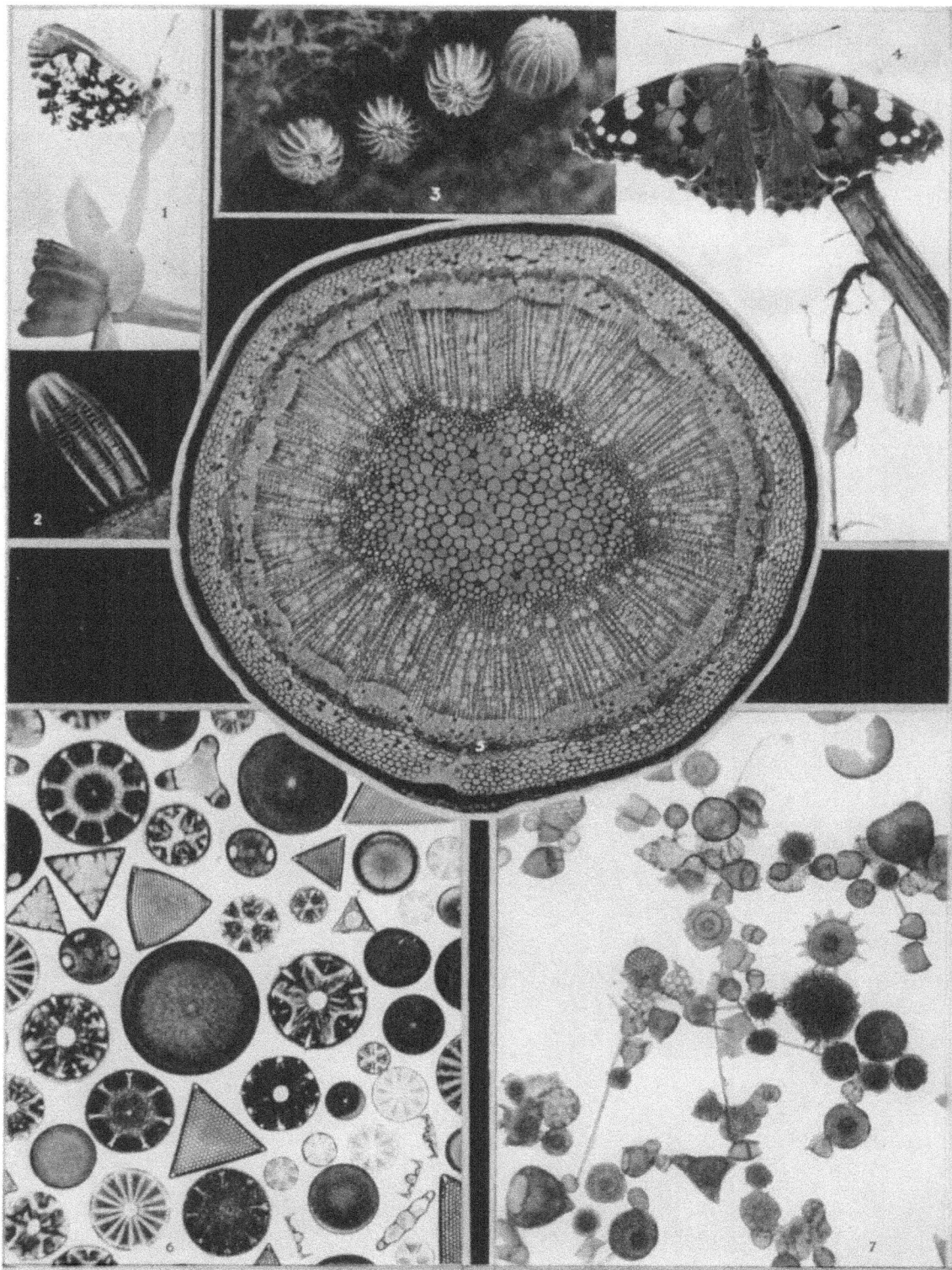

WONDERS OF THE MICROSCOPE

1. Male Orange-tip Butterfly (*Euchloe cardamines*) uncoiling its proboscis. 2. Egg of Orange-tip Butterfly, magnified 25 diameters. 3. Eggs of Painted Lady Butterfly, magnified 25 diameters. 4. Painted Lady Butterfly (*Pyrameis cardui*), just emerged from its chrysalis; the broken pupa skin is seen beneath. 5. Transverse section of young twig of Beech (*Fagus sylvatica*) with tissues artificially stained to aid in their identification, e.g. innermost the pith (green), the wood (blue), magnified 40 diameters. 6. Siliceous shells which enclose the microscopic plants called Diatoms, magnified 100 diameters. 7. Siliceous skeletons of the unicellular animals known as Radiolaria, magnified 100 diameters.

Chromo-photos: J. J. Ward, F.E.S.

THE
OUTLINE OF SCIENCE

A PLAIN STORY SIMPLY TOLD

EDITED BY

J. ARTHUR THOMSON

REGIUS PROFESSOR OF NATURAL HISTORY IN THE
UNIVERSITY OF ABERDEEN

WITH 40 COLOURED PLATES

AND

800 ILLUSTRATIONS IN BLACK AND WHITE

IN FOUR VOLUMES

★ ★

WILDSIDE PRESS

www.wildsidepress.com

CONTENTS

ILLUSTRATIONS

vi   **Illustrations**

Illustrations ix

FACING PAGE

MENDEL'S LAW ILLUSTRATED IN PEAS . 380

Illustrations ix

Illustrations ix

Illustrations ix

Illustrations ix

Illustrations ix

FACING PAGE

Illustrations ix

FACING PAGE

Illustrations ix

FACING PAGE

MENDEL'S LAW ILLUSTRATED IN PEAS 380

MENDELIAN INHERITANCE IN WHEAT 381
(After R. H. Biffen.)

THE IDEA OF GERMINAL CONTINUITY 381

MENDELISM IN MICE 384
Photo: British Museum (Natural History).

MENDELIAN INHERITANCE IN ANDALUSIAN FOWLS 385
(After Darbishire.)

HALF-LOP RABBIT 388

INHERITANCE IN SNAILS, WITH BANDLESS AND BANDED VARIETIES . 388
(After Lang.)

THE RUFF (*Machetes pugnax*) 389
(From specimens.)

MODEL OF THE EXTINCT DODO 398
Photo: Royal Scottish Museum.

THE EVOLUTION OF BIRDS 398
After W. P. Pycraft.

BIRDS AND THEIR FOOD 399

GREAT AUK (WITH EGG) 402
Photo: Royal Scottish Museum.

THE KIWI OR APTERYX 402
Photo: James's Press Agency.

GANNET—A LESSON FOR AVIATORS 403
From *Our Common Sea-Birds* (Country Life, Ltd.)

THE GANNET—PUTTING ON THE BRAKE 403
From *Our Common Sea-Birds* (Country Life, Ltd.)

HERRING-GULLS (*Larus argentatus*) 406
From *Our Common Sea-Birds* (Country Life, Ltd.). Photo: J. C. Douglas.

GOLDEN EAGLE 406
Photo: Royal Scottish Museum.

RAVEN 407
Photo: James's Press Agency.

SECRETARY BIRD 407
Photo: F. W. Bond.

Illustrations

The Outline of Science

IX
THE WONDERS OF MICROSCOPY

THE WONDERS OF MICROSCOPY

THE use of a lens for magnifying purposes is ancient, but the first "compound microscope" was probably made in 1590 by a Dutchman, Zacharias Jansen, whose invention was followed up by Galileo a few years later. But it did not become an effective instrument till towards the middle of the eighteenth century. In a "simple" microscope we look at the object directly through a lens or through several lenses. This kind of instrument is often used for microscopic dissection. But in the "compound" microscope we look through an eye-lens or ocular at an inverted image of the object formed inside the tube of the instrument by an object-lens or objective. In all ordinary microscopes there are two lenses in the eye-piece and three lenses in the objective, and all sorts of ingenious devices have been invented for making the most of the magnifying power without losing clearness and definition.

An Invisible World of Life

In the early days of microscopy the instrument was to a large extent a scientific toy. The observers magnified objects and often drew them very beautifully, but without making them more intelligible. There is not much gain in seeing a minute object loom large unless we understand it better. This was a necessary stage. Soon, however, great steps were taken, and one of these may be called *the discovery of the invisible world of life.* The pioneer explorer was surely the Dutch observer Leeuwenhoek (1632-1723), who discovered minute creatures like the Rotifers or Wheel-Animalcules which are common in ponds, and the Infusorians which abound wherever vegetable matter rots away in water. He made numerous microscopes, and though they had neither tube nor mirror, they were sufficient to enable him to demonstrate his animalcules before the Royal Society of London, the Fellows signing an affidavit that they had seen the little creatures. It was Leeuwenhoek also who (in 1687) discovered bacteria, the very minute organisms which cause all putrefaction, are responsible for bringing about many diseases, and are yet of immense service to many living creatures.

It was not till long afterwards that Pasteur and others demonstrated the importance of bacteria, but it was a great event in the history of science when Leeuwenhoek first proved their presence. It was literally the discovery of a new world with a teeming population, with incalculable powers for good and evil. It must have been a seed in the human mind, this idea of an intense activity going on all unseen until men stuck lenses of glass in front of their own.

Another great event, though its importance was not recognised till afterwards, was *the discovery of the male elements* or spermatozoa of animals, which fertilise the egg-cells so that these may begin to develop. This discovery was probably due (1677) to a medical student in Leyden, Louis de Hamen, who showed them to Leeuwenhoek, but it was not till more than a hundred years later that the meaning of these sperm-cells was recognised. And it is interesting to remember that it was not till 1843 that another medical student, Martin Barry, in Edinburgh, observed for the first time in the rabbit the fertilisation of the mammalian ovum by the spermatozoon. In modern times an extraordinary intensity of research has been focused on the usually microscopic egg-cell and the always microscopic sperm-cell. In the union of these an individual animal has its beginning and it is interesting

FULL-FACE PORTRAIT OF THE COMMON WASP

The large compound eyes are well seen, and above these are the feelers or antennæ. Protruding underneath the front of the head are some of the biting mouth-parts, and the first pair of legs are also seen. Note also the numerous setæ or bristles.

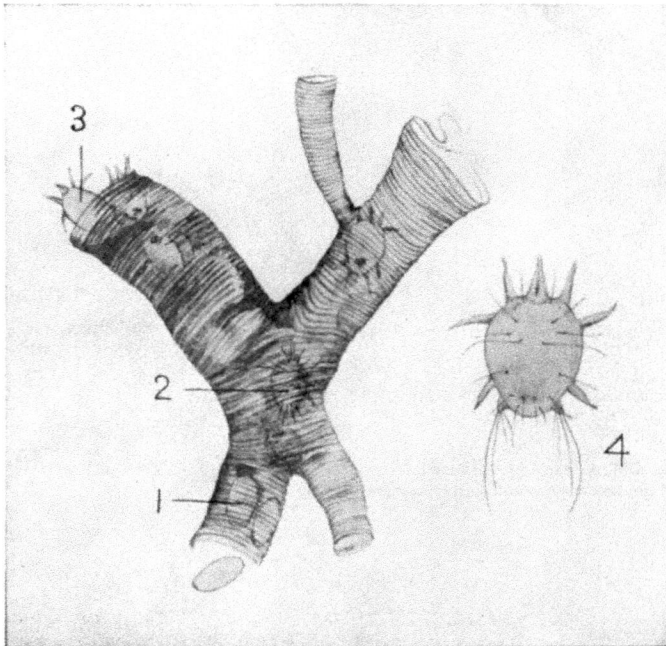

"ISLE OF WIGHT DISEASE" HAS SPREAD ALL OVER BRITAIN, CAUS-ING THE RUIN OF THOUSANDS OF BEEHIVES

It is associated with the presence of a very minute mite (*Tarsonemnus woodi*) in certain air-tubes or tracheæ of the bee. The figure shows an enlargement of a branching air-tube with a mite's egg (1), an immature female (2), a mature female (3) struggling out. In Fig. 4 is shown a mature mite still more en-larged; note the four pairs of walking-legs with bristles and two pairs of piercing and sucking mouth-parts. The mite feeds on the bee's blood, and as the num-bers increase the infected air-tubes become blocked. This means that certain muscles, e.g. those of flight, are bereft of their normal supply of oxygen and naturally go out of gear. The bees are unable to fly, and crawl about help-lessly in front of the hive.

EGGS OF THE BROAD-BORDERED UNDERWING MOTH, CROWDED TOGETHER ON THE UNDERSIDE OF THE LEAF

There are about 1,300 eggs, all deposited in one evening. In another photograph a few of the eggs are shown much enlarged.

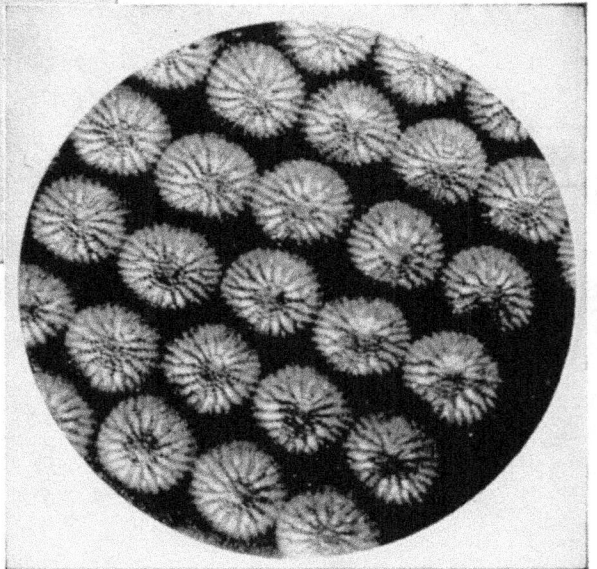

EGGS OF THE BROAD-BORDERED UNDERWING MOTH

Each egg has a shell of the substance called chitin, which covers the living skin of all insects. The egg-shell is secreted in the oviduct of the female insect. Its outer surface is very beautifully sculptured, but the significance of the pattern is unknown. It looks as if the beauty of many organic structures is like that of snow crystals—an expression of the way in which the dance of molecules sinks into relative rest.

to trace this modern study, so important in connection with heredity, back and back to the Leyden student's first glimpse of spermatozoa.

But we must not lose the wood in the trees: one of the real wonders of microscopy, rising high above any mere curiosity collecting, is the discovery of a world of invisible life. There are the bacteria, which may be regarded as the simplest of living creatures; there are the yeasts and the simple moulds; there are the single-celled green plants which play so important a rôle in the economy of the sea by providing food for humble animals like water-fleas. There are the one-celled animals or Protozoa, such as the chalk-forming Foraminifera, the Infusorians which often serve as middlemen between the products of bacterial putrefaction and some higher incarnation in crustacean or worm, and the death-bringing organisms of malaria and sleeping-sickness. There are also many-celled animals of miscroscopic dimensions, such as the wheel-animalcules of the pond and the minute crustaceans which play so important a part in the circulation of matter by feeding on the microscopic Algæ and Infusorians in the water and being themselves devoured by fishes. There are also the invisible early stages of many important parasites, whose life-history would have remained quite obscure if naturalists had been without micro-scopes. It seems hardly too much to say that the system of animate nature would be uncomfortably magical if the microscope had not enabled us to detect the missing links in many a chain of events. The liver-fluke which often destroys the farmer's sheep is a relatively large animal—about an inch long—but it starts its life as a microscopic egg which develops into a microscopic larva that enters a water-snail, and has a remarkable history there. The tapeworm, with which man becomes infected by eating bad beef imperfectly cooked, may be several yards in length, but it began as a microscopic egg which was swallowed by a bullock and hatched into a microscopic boring larva, which eventually became the beef-bladderworm. In hundreds of cases the microscope reveals the life-history. In the course of a few years a very serious bee-pest, known as Isle of Wight disease, has spread throughout Britain, causing havoc among the hives, and greatly discouraging a lucrative and wholesome industry. The nature and meaning of this disease remained baffling until strenuous and patient microscopic work by Rennie and White demonstrated that the plague was bound up with the presence of an extremely minute mite in the anterior breathing-tubes of the bee. And when the cause of a disease is discovered, it is not usually long before investigation also reveals a cure.

Intricacy of Architecture in Small Animals

Long before there was any microscope the use of the scalpel, helped sometimes by the simple lens, had revealed the intricacies of the body in man and in animals. We may save ourselves from exaggerating modern achievements by recalling how much Aristotle (384-322 B.C.) knew of animal structure. He dissected many a creature, such as the sea-urchin; he saw the beating of the tiny heart of the unhatched chick; he described how the embryo of the smooth dogfish is bound to the wall of its mother's oviduct, and much more besides. And Aristotle had his successors, few and far between, who kept up the anatomising tradition, long before there was any microscope. But what the early microscopists did was to reveal the fact that the multitude of minute creatures which it was hopeless to try to dissect had an intricacy of structure comparable to that in larger and higher animals. One of the pioneers in this exploration was the Italian, Marcello Malpighi (1628-1694), who described the internal architecture of the silkworm as animal had never been described before. He worked so hard that he threw himself into a fever and set up inflammation in his eyes. "Nevertheless, in performing these researches so many marvels of nature were spread before my eyes that I experienced an internal pleasure that my pen could not describe." He discovered, for instance, the delicate branching air-tubes (or

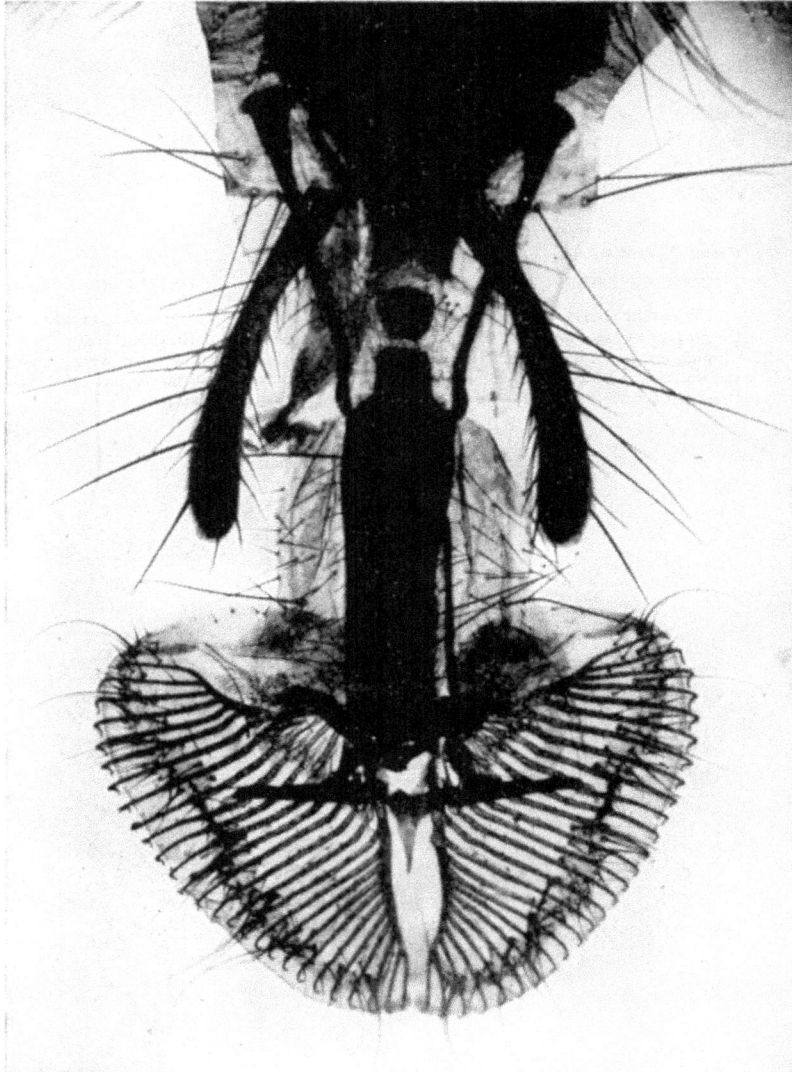

THE PROBOSCIS OF A HOUSE-FLY

It is a tubular structure ending in two broad pads, traversed by numerous small canals by which saliva from the mouth can be passed out on the surface of the pads.   This juice dissolves solid substances like sugar, and the result is sucked up the proboscis.   Two unjointed sensitive palps are also shown.   When not in use the proboscis is drawn into a hollow beneath the fly's head.

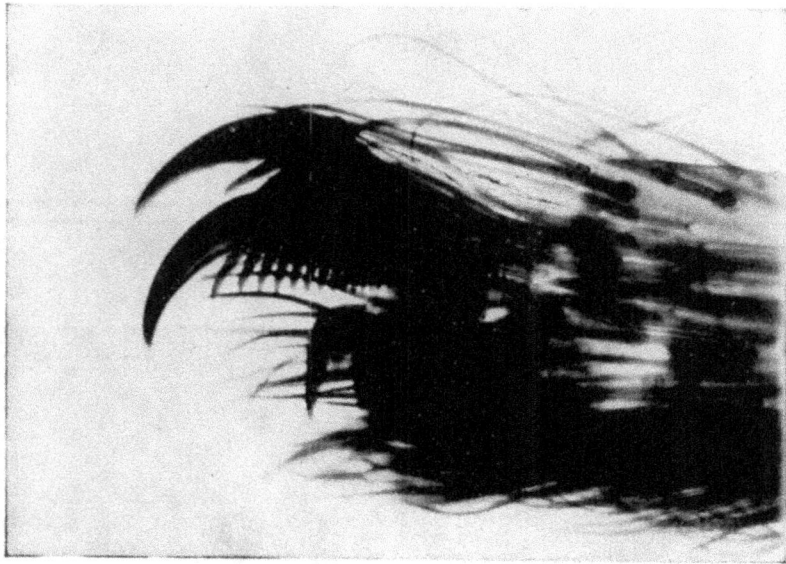

FOOT OF SPIDER, SHOWING COMBED CLAWS AND CUTTING-HOOK BELOW

When the spider runs up a wall or creeps along a ceiling, it is gripping the roughnesses on the surface by means of these combed claws. There is mention in the Scriptures of the spider "laying hold with her hands"; but some scholars say that the Hebrew word really refers to the wall-lizard called the Gecko, which has toes with adhesive surfaces. A spider cannot climb up a quartz fibre, for its claws will not grip the smooth surface.

FOOT OF A WINGLESS FLY, *Melophagus Ovinus*, OFTEN, BUT BADLY, NAMED THE SHEEP-TICK

For a tick is not an insect at all. The "ked," as it is popularly called, has a compressed body about a quarter of an inch long. It has piercing mouth-parts and sucks blood from the sheep. Very striking are the two curved claws at the tip of the foot, well suited for holding on to the fleece. Part of the body, with short hairs, is also shown. The "keds" usually pass from one sheep to another by contact.

tracheæ) which carry air to every hole and corner of the insect's body; and it is plain from this instance that he discovered internal structures which made the insect at once more intelligible. This sort of discovery (we still call the excretory organs of insects Malpighian tubes) was characteristic of the man, and characteristic of a kind of investigation which continues untiringly to the present day. It makes for a realisation of the unity of organic nature to disclose in creatures which will pass through the eye of a needle the presence of organs comparable to those in man himself. Much of Malpighi's work was done with a simple lens, but he had also his microscope with two lenses, and in any case his name may be associated with the great discovery that, as far as intricacy of structure goes, size does not count for much.

It is a very striking experience to observe a minute animal like the Rotifer Hydatina, no more than a pin-prick in size, and to find that it has a food-canal, a chewing apparatus, a nerve-centre, various muscles, a delicate kidney-tube, and so on. Yet it is such a pigmy when all is said. There are little beetles (Trichopterygids), well represented in Britain, which are sometimes only one-hundredth of an inch in length—practically invisible. Yet within that small compass there is the same kind of intricacy that is found in a Goliath Beetle—brain and nerves, muscles and food-canal, air-tubes and kidney-tubes, blood and germ-cells. He would be a bold man who says he quite understands how there is all this intricacy within bulk so small. But this we venture to call the second wonder of microscopy, that *great intricacy of structure may occur in a microscopically minute living body.*

Intricacy of Vital Architecture

We have singled out the name of Malpighi in Italy as a pioneer in the exploration of the structure of minute animals, but we might have taken with equal justice Swammerdam in Holland, whose precision of minutiose observation has rarely been equalled.

He is memorable not only for his anatomy of small creatures, but, like Malpighi, for his minute anatomy of larger ones, and here we might also include the early British microscopists, Hooke and Grew. For this was another line of advance, to disclose the intricacy of vital architecture that lay beyond the limits of scalpel and simple lens. Thus it was a great step when Swammerdam discovered in 1658 the blood corpuscles of the frog; when Malpighi demonstrated the air-cells in the lung where the gaseous interchange takes place between blood and air; when Leeuwenhoek completed Harvey's theory of the circulation of the blood by demonstrating in 1680 the capillary connection between arteries and veins. Speaking of the tail of the tadpole, he said, "A sight presented itself more delightful than any mine eyes had ever beheld; for here I discovered more than fifty circulations of the blood in different places, while the animal lay quiet in the water, and I could bring it before my microscope to my wish. For I saw not only that in many places the blood was conveyed through exceedingly minute vessels, from the middle of the tail toward the edges, but that each of the vessels had a curve or turning, and carried the blood back toward the middle of the tail, in order to be again conveyed to the heart." Such was the momentous observation of the fact that the arteries leading *from* the heart, and the veins leading *back to* the heart are bound into one system by the intermediation of the capillaries.

This is an easy illustration of the kind of service microscopy has never ceased to render—making vital activity more intelligible by revealing the intricacy of structure. For it is in a study of the structure that we get a better understanding of the ways and means of life. It is not the whole story of the workshop to know the furnishings and the tools, but it is an essential part of the story. We hastily draw away our finger from a hot plate—a reflex action —it is only with the help of the microscope that the physiologist can tell how the message travels by sensory nerve-cells to intermediary nerve-cells and thence to motor nerve-cells which com-

Photo: J. J. Ward.

COILED PROBOSCIS OF A BUTTERFLY

It is a paired structure, but only one-half of the tube is shown.   It is used for sucking up nectar from flowers, the suction being due to a muscular pump inside the insect's head.   In some of the Hawk Moths the proboscis may be 10 inches long.   Rising from near the base of the proboscis there are two palps covered with sensitive hairs, as the photograph shows.   The caterpillar or larva of the butterfly has biting mouth-parts in great contrast to the suctorial parts of the adult.

ORGANISM
ORGANS
TISSUES
CELLS
PROTOPLASM

A living creature or *organism* is usually built up of *organs*, such as heart and kidneys; the organs are composed of *tissues*, like muscular and nervous, glandular and connective tissue; the tissues, e.g. a piece of flesh, are built up of microscopic unit-corpuscles or *cells* of various kinds; and the cells consist largely of living matter or *protoplasm*.

Photo: J. J. Ward.

RASPING RIBBON OR FILE, BADLY CALLED THE
PALATE, IN THE MOUTH OF THE WHELK

By means of this toothed flexible file the whelk can bore a hole through the skin, or even through the shell, of an animal on which it preys.

JOHN GOODSIR, 1814–1867, PROFESSOR OF ANATOMY
IN THE UNIVERSITY OF EDINBURGH

A remarkable pioneer who had an important share, along with Schwann, Schleiden, and Virchow, in establishing the Cell-Theory—one of the foundation-stones of Biology.

mand the muscles to move. Our mouth waters at the sight of palatable food: it is only by help of the microscope that the physiologist is able to trace the message from eye to salivary glands, and to show how in the cells or unit-corpuscles of these glands there is a preparation of secretion which is discharged when the trigger is pulled by a nervous command. The study of vital activity requires experiment and chemical analysis, but it cannot dispense with the microscope. So we venture to say that a third wonder of microscopy is the revelation of *the intricacy of minute structure*.

The Stones and Mortar of the House of Life

It is a long and tangled story which tells of the gradual discovery of the cells or unit-areas of which all but the simplest living creatures are built up, and of the living matter or protoplasm which these cells contain or portion off. The genius of the short-lived French anatomist Bichat had analysed a living body into a web of tissues—nervous, muscular, glandular, connective, and epithelial. But to Schwann and Schleiden, Virchow and Goodsir, is due the credit of a further advance—the Cell-Theory—certainly one of the triumphs of microscopes with brains behind them. The Cell-Theory or Cell-Doctrine states three facts: (1) that all plants and animals have a cellular structure (being either single cells or combinations of numerous cells); (2) that every living creature, reproduced in the ordinary way, begins its life as a single cell, and, if it does not remain at that humble level, proceeds to build up a body by the division and re-division of cells which eventually form tissues and organs; and (3) that the activities of a many-celled organism are the co-ordinated summation of the activities of the component cells. "Every animal," Virchow said, "appears as a sum of vital units." Not that we are to think of an ordinary animal as a colony of cells, as a mob is a collection of angry men, or even as a battalion is a co-ordination of disciplined soldiers. It is nearer the truth to think of the fertilised

egg-cell—a *potential organism*, when we come to think of it—
dividing and re-dividing into cells so that the unified business of
life may be more effectively carried on by division of labour. As
one of the greatest of botanists said: It is not that the cells form
the plant; it is rather that the plant makes the cells.

The Microcosm of the Cell

With a few exceptions, notably Aristotle, the early naturalists were content to study the outsides of animals; then came
the study of internal organs, like hearts and lungs; Bichât marks
the deeper penetration to the tissues that make up the organs;
then came the recognition of the cells that compose the tissues;
finally there was the recognition of protoplasm—which Huxley
called "the physical basis of life." It may be useful to place the
different levels of study in a clear scheme (see illustration
facing p. 305).

The old picture of a cell was that of a little drop of living
matter with a kernel or nucleus, and sometimes with an enclosing
wall. But the revelations of the microscope have made this
picture obsolete. We have to think of a more or less unified
minute area of great chemical diversity, with complex particles
and unmixing droplets restlessly moving in a fluid. In the centre
of this whirlpool, with its flotsam of reserve-products and waste-
products, there floats the nucleus, a little world in itself. Inside
its membrane, through which materials are ever permeating out
and in, there are readily stainable nuclear bodies or "chromosomes," usually a definite number for each species. And each
"chromosome" is built up of bead-like "microsomes" strung on a
transparent ribbon. It begins to make one's head reel—cell,
nucleus, chromosomes, microsomes! But it is all fact.

Inside the nucleus there may be a nucleolus or more than one,
and outside the nucleus there is a minute body called the centro-
some, which plays an important part in the division of the cell.
This is not nearly all, but it is enough to suggest *how complex is*

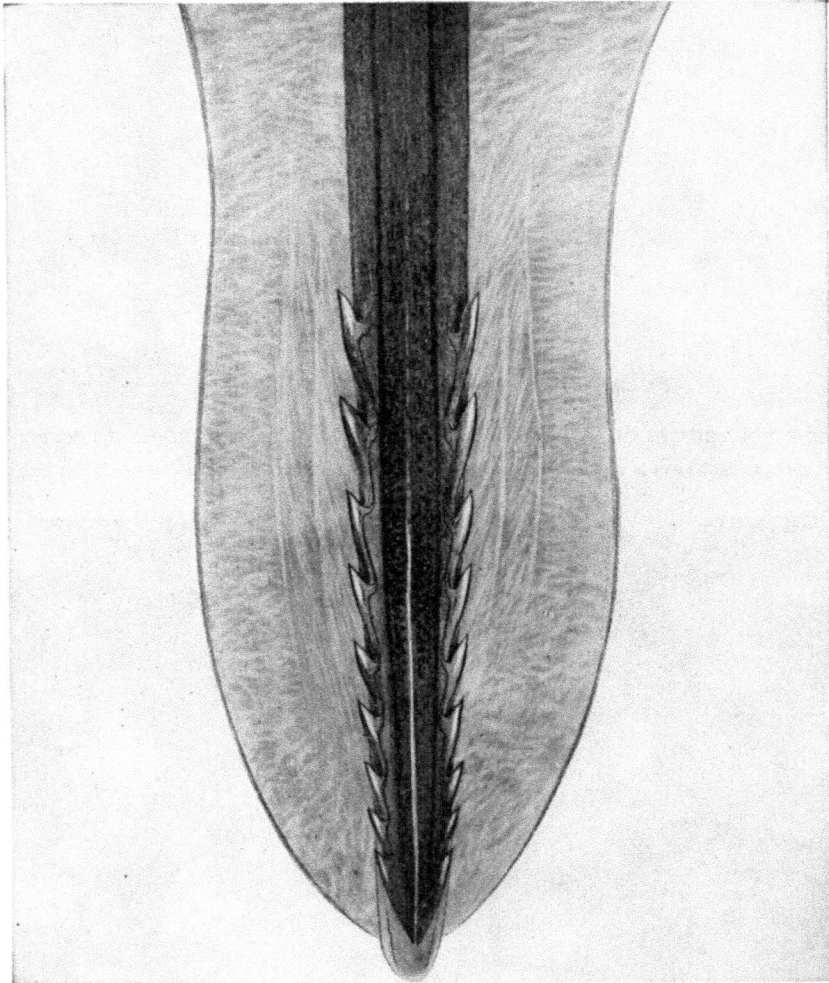

STING OF THE HONEY-BEE

In the middle there is a deeply grooved "guide," pointed and barbed at the tip. Working in a concavity on each side of the "guide," is a slender lancet or "dart," with recurved barbs at the tip. There are beadings which keep the darts sliding smoothly along the guide. The tips of the darts can be protruded beyond the end of the guide. Enclosing the three dark-coloured piercing structures there are two fleshy "sting-palps," shown in lighter colour in the figure, which includes only the *terminal* part of the sting. When the bee strikes, the secretion of the poison-gland **is** forced down between the darts and the guide.

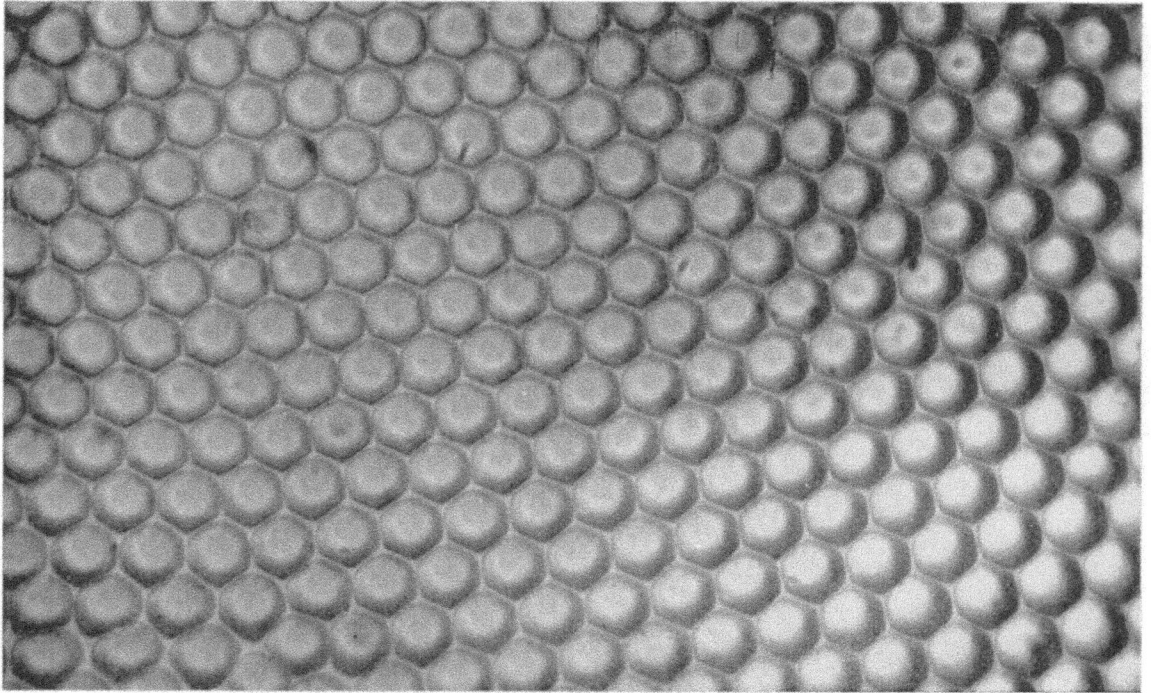

SURFACE VIEW OF THE COMPOUND EYE OF A FLY, SHOWING SOME OF THE THOUSANDS OF CORNEAL FACETS

Corresponding to each facet is a complete "eye-element," including a double lens and a percipient retina or retinule.

PHOTO OF STATUE TAKEN THROUGH THE LENSES OF THE FLY'S CORNEA

A multitude of images has been formed, one image through each of the thousands of lenses. The photo-micrographs of the fly's eye were taken by means of Davidson's "Davon" Patent Super-Microscope, in which an achromatic combination called a "collector" is placed behind the objective in the microscope. This projects an "air" image of the object under examination beyond it, and this "air" image is magnified by another objective and eyepiece acting as a compound eyepiece. The result is that higher magnification can be employed on any given objective than is usually employed and without loss of resolution.

the microcosm of the cell. Inside each of man's cells there are about two dozen chromosomes, and one of the authorities on cell-lore speaks of each chromosome having the corporate individuality of a regiment, the really indivisible living units being the beads or microsomes—which correspond to the men! And to this must of course be added the fact that we have many millions of these cells in our body. Indeed, we are fearfully and wonderfully made!

The Beginning of the Individual

Every many-celled creature, which reproduces in the ordinary way, starts on the journey of life as a single cell—the fertilised ovum. As we have made clear in a previous article, the usually microscopic fertilised egg-cell contains, in some way that we cannot picture, the initiatives or "factors" for the hereditary characters of the living creature in question. But the microscope has begun to reveal the little world within the egg-cell, and it has been found possible to map out the way in which the factors for certain characters are disposed in the chromosomes. Thus in the case of the egg of the fruit-fly called Drosophila, it is possible to say that the hereditary or germinal factor for, say, red eye or grey wing, lies at such and such a level in one of the four chromosomes. *It would be difficult to find a wonder of microscopy greater than this.*

Yet this is but an instance of what goes on at a level of visibility which only the microscope can reach. We know much in regard to the permutations and combinations which take place when the germ-cell is ripening—shufflings of the hereditary cards which throw some light on the origin of new departures. We know something of the manner in which the paternal and maternal hereditary contributions behave in relation to one another when and after fertilisation takes place. We know much in regard to the sequence of events in individual development, wherein the obviously complex emerges from the apparently simple, and the implicit inheritance becomes an explicit individual. In the seventeenth century, William Harvey, the discoverer of the circulation of the blood, wrote in regard to development: "Although it be a known thing subscribed by all, that the fœtus assumes its original and birth from the male and female, and consequently that the egge is produced by the cock and henne, and the chicken out of the egge, yet neither the schools of physicians nor Aristotle's discerning brain have disclosed the manner how the cock and its seed doth mint and coine the chicken out of the egge." But although we do not understand to-day how the factors of an inheritance are condensed into the dimensions of a pin-prick; or how the fertilised egg-cell segments into two, and cleavage after cleavage continues, with associated division of labour, until an embryo is built up; we *do* know why it is that like tends to beget like, why certain hereditary characters are distributed in a particular way among the offspring. And we also know the successive steps by which the process of development is accomplished. It is this kind of knowledge, we think, which must be regarded as *the crowning wonder of microscopy*. These fundamental questions of heredity and development will be discussed in a separate article, but the point here is that the scientific study of inheritance can as little dispense with microscopy as with breeding experiments and statistics. All three are *essential*.

Manifold Uses of the Microscope

Everyone knows that finger-prints are sometimes of critical importance in the identification of a criminal. The details of the pattern of the ridges on the fingers vary from man to man; they are *individual*. Therefore, if a good impression is available on some surface which has been handled in the course of a burglary, let us say, it can be compared with the collection in the album of criminals' finger-prints, and identification may follow. The microscope has even subtler use in *the detection of crime*. If splashes of blood on the clothes of a suspected murderer are declared by him to be due to the blood of a rabbit which he killed, it is usually possible to test the truth of his statement *microscopically*. For the

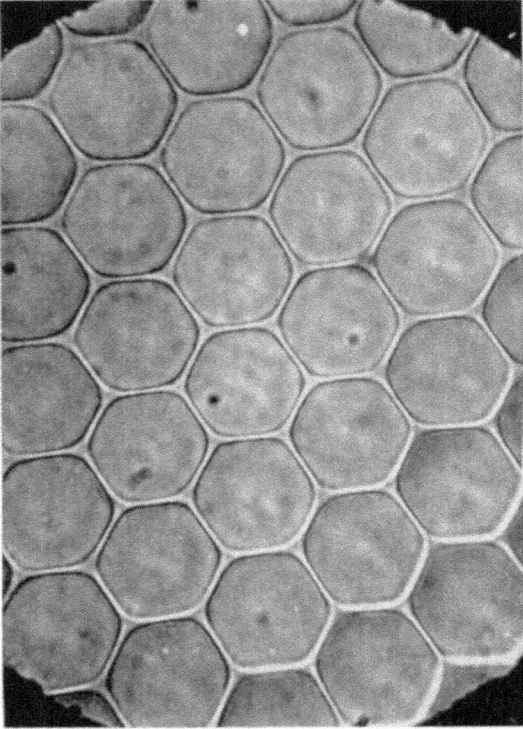

SURFACE VIEW OF THE COMPOUND EYE OF A FLY,
SHOWING THE NUMEROUS HEXAGONAL FACETS ON
THE TRANSPARENT CORNEA

To the inside of the facet there is an outer or corneal lens,
inside that a crystalline cone or inner lens, and inside that
again a percipient rod.

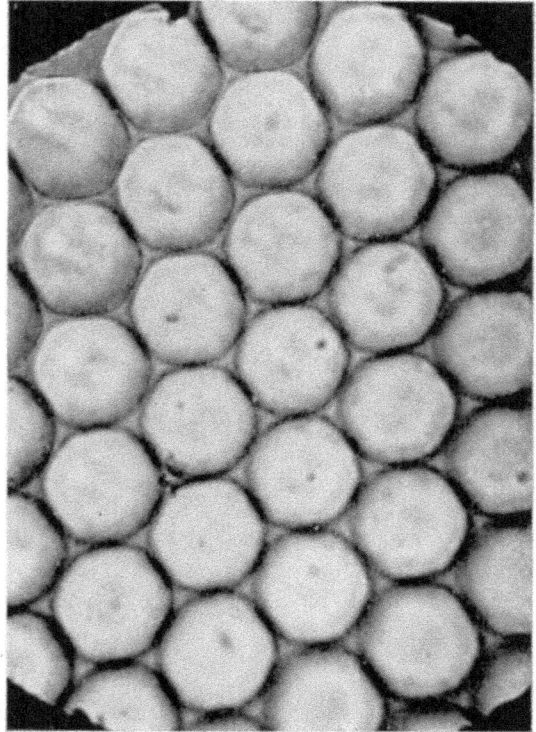

SURFACE VIEW OF THE SAME EYE, SHOWING THE
CORNEAL OR OUTER LENSES FITTING AGAINST THE
HEXAGONAL FACETS OR FRAMES

In the house-fly there are about 4,000 facets, corneal lenses,
crystalline cones, and sensitive rods.

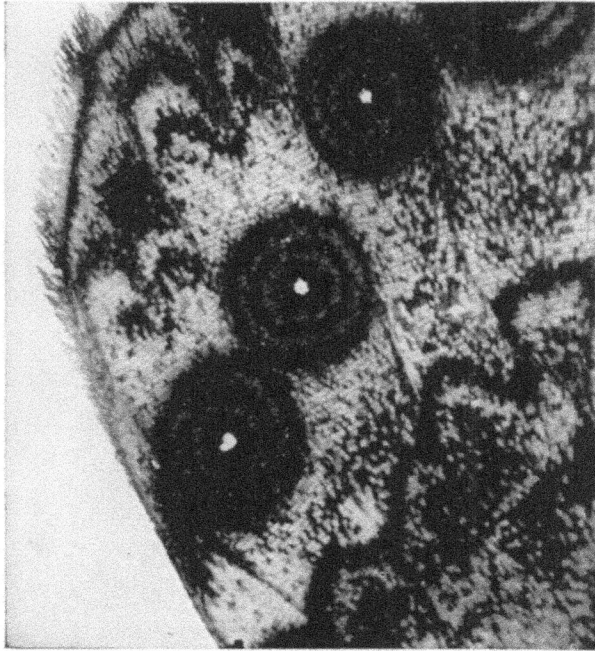

PART OF THE WING OF A WALL-BUTTERFLY

Showing how the numerous scales build up its design, not only covering the general surface like the slates on a roof, but arranged in wavy
bands and concentric zones. Each microscopic scale is finely striated
longitudinally, and the light falling on these undergoes "interference."
This gives the metallic iridescence to the wings of many butterflies,
greatly enhancing the coloration due to pigment.

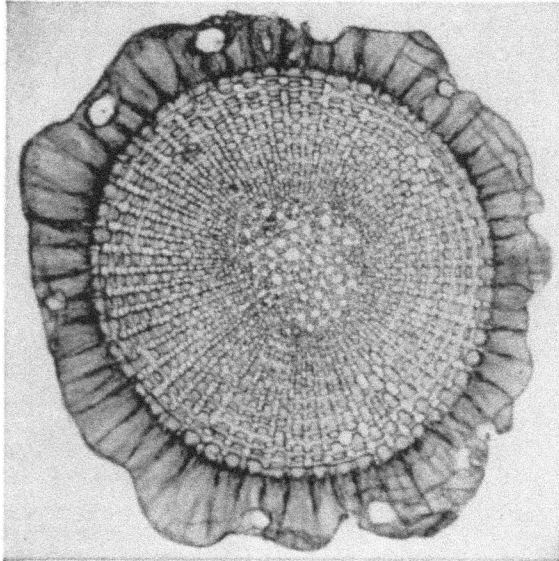

CROSS-SECTION OF THE SPINE OF A SEA-URCHIN (*Cidaris
Metularia*), MAGNIFIED SEVERAL DIAMETERS

The remarkable feature is the beautiful zoned structure. The
spine is a delicate needle to start with, but around this there is
a periodic deposition of lime from the skin, one concentric zone after
another. It is probable that this kind of architecture gives the
spine great stability, but it is primarily an expression of rhythmic
orderly growth. It shows that the quality of beauty is not confined
to the outsides of animals.

dimensions of the very minute red blood corpuscles differ in different mammals, and the *circular shape* in all mammals (except camels) can be distinguished at a glance from the *elliptical* shape in all the other backboned animals. Moreover, the red blood pigment of hæmoglobin can be easily made to assume a crystalline form, and it is a very remarkable fact that the blood-crystals of the horse can be distinguished microscopically from those of the ass, and even those of the domestic dog from those of the wild Australian dingo. Poisons that crystallise may also be detected by means of the microscope.

The use of the microscope in *medicine* may be illustrated in reference to the blood. For it is often possible by microscopically examining a film of blood, spread on a slide, to tell what is wrong with the patient. Microscopic parasites may be detected, like those of malaria; methods of counting the red blood corpuscles (man has trillions!) may show that they are far below the proper number; and a change in the normal shape of the hæmoglobin crystals may show that something is amiss. It is unnecessary to dwell on the medical importance of the microscope in determining the presence or absence of certain kinds of microbes and higher parasites in the blood or food-canal of the patient. Along with this physiological utilisation of the microscope we may take its use in testing drinking water, which is liable to be fouled by the presence of bacteria and various minute animals. Also of great importance is the microscopic study of milk, for this fluid is peculiarly liable to contamination, and is very suitable for the growth of various kinds of disease germs.

For *the detection of adulteration* the microscope is also invaluable. The starch grains of different plants, such as potato, wheat, rice, maize, are readily distinguished from one another, and a microscopic examination may immediately prove that a commodity sold under a particular name, e.g. as arrowroot, is not what it professes to be. If a sample of so-called "honey" contains no pollen-grains, but a great many starch-grains, we may be sure that the busy bee was not the chief agent in its production. In short, the microscope is a valuable detective of dishonesty. But a use of the microscope more important and more pleasant to think of is in metallurgy, where its utilisation to detect the structural features of the stable and the transient in various metallurgical combinations, such as different kinds of steel, has been of inestimable importance.

A farmer can always make good use of a lens in examining samples of seeds, or in identifying particular kinds of injurious insects, or in detecting the beginnings of "rusts" and "mildews" on his crops. But the expert agriculturist must of course go much further, especially in warm countries, where the microscope is necessary for the study of the insidious Fungi which are always ready to find a weak spot in the plants' defences—in all sorts of plantations from coffee to rubber.

The Ultra-Microscope

Early in the twentieth century an ingenious method was described by Siedentopf and Zsigmondy, which is often briefly referred to as the *ultra-microscope*. Everyone knows from personal observation that a strong beam of sunlight entering a darkened room reveals a multitude of dust particles, which are not seen at all in ordinary light. The same multitude of particles is often seen in the track of a strong beam from a "magic lantern" in a darkened room. These dancing particles, whose abundance we scarcely suspected, become visible because they are so strongly illumined; there is a diffraction of rays from their surface, and they look much bigger than they really are.

In 1899 Lord Rayleigh pointed out that a particle too small to be seen by the highest power of the microscope under ordinary conditions might be made visible if it received sufficiently intense illumination; and the ultra-microscope took advantage of this idea. It occurred to Siedentopf and Zsigmondy that if the particles in a solution could be strongly illumined by a beam coming

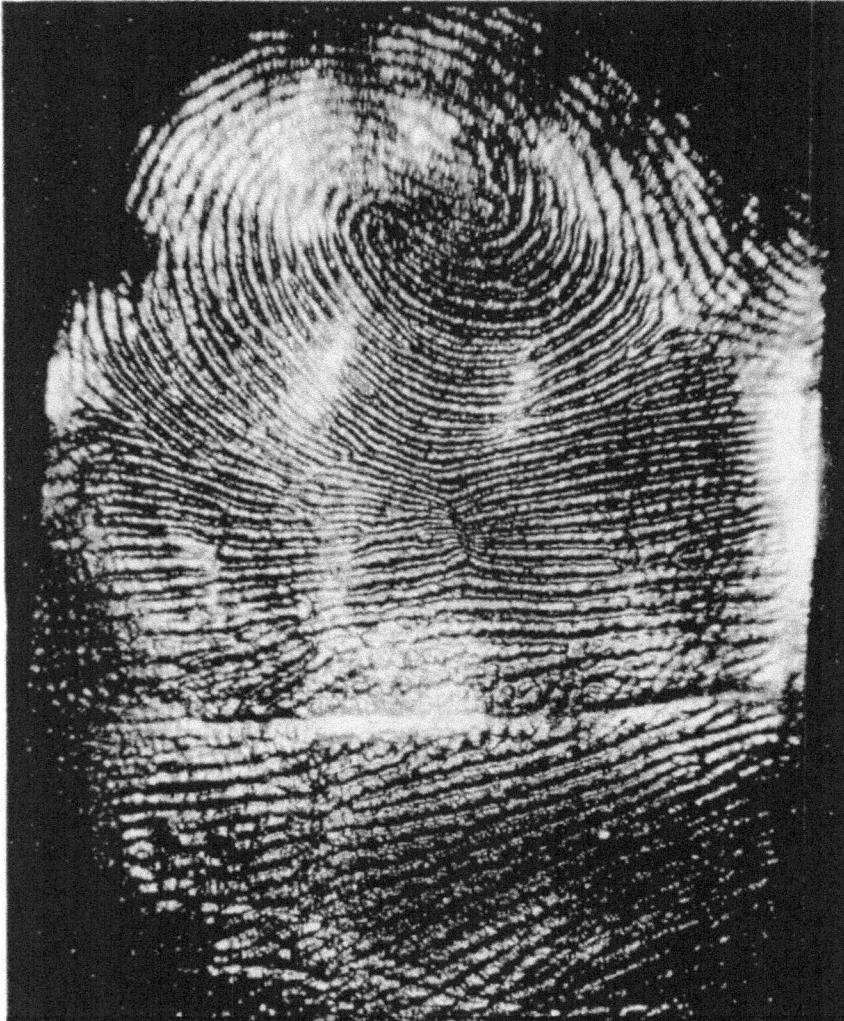

THUMB-PRINT IN WAX

The skin is marked by numerous ridges and intervening valleys, which have individual peculiarities of pattern, even in the same family. This illustrates inborn variability. The individuality of the pattern is so marked that the prints are used as sure means of identification.

Photo: J. J. Ward.

FEET OF THE JUMPING-LEGS OF THE COMMON FLEA

The powerful muscles which give the flea its familiar power of leaping are mainly in the uppermost part of the leg, which is not shown here; but there are minor muscles continued to the very tip of the legs, to the base of the terminal claws. Note the numerous bristles or setæ on the leg.

in, so to speak, sideways, then particles ordinarily invisible might stand out. Their diffraction-images, at any rate, would be seen. In ordinary microscopic conditions the beam of light is thrown by the mirror, usually through a sub-stage condenser, directly through the solution or thin transparent section, up into the tube of the microscope, where an image is formed, to be re-formed by the eyepiece. In the ultra-microscope for examining solutions the beam of light is projected *horizontally* into the solution and examined from above. The result is that particles ordinarily invisible are seen in a vigorous dance, the so-called Brownian movement. This dance is due to the particles being bombarded by the moving molecules of the fluid in which they are suspended. By accessory devices it becomes possible, in the use of the ultra-microscope, to count the number of particles in a solution and to measure the mass of each. This has formed the basis of exceedingly interesting conclusions which are unfortunately beyond our scope in this article.

A reference should be added, however, to another method called "dark-ground illumination" which makes structures visible which are invisible in ordinary conditions of microscopic work. Professor Bayliss writes: "The central rays of the illuminating beam are cut out by means of a stop, and the peripheral rays are reflected by a parabolic surface so as to meet in a point in the object under examination; they cross at such an angle as to pass outside the field of the objective in use, which only picks up the light refracted, or diffracted, from structures in the preparation." The dark-ground illumination brings out features which are invisible in the ordinary direct illumination.

The essential parts of a microscope are, as we have seen, (1) the objective for obtaining the first magnified image of the object; (2) the ocular for further enlarging that image and transmitting it to the observer's eye; and (3) the sub-stage condenser for illuminating the object with a cone of light. Now, in modern times, there have been numerous detailed improvements in these parts, e.g. in the quality of the glass used in making the lenses; and a present-day microscope is certainly a very perfect instrument. Indeed, unless some new idea is discovered, such as those behind the ultra-microscope and dark-ground illumination, it does not seem likely that great advances in technical microscopy can be made. The reason for this statement is to be found in the optical limitations of the instrument. The use of the microscope is not mainly magnification but *resolution*. "By resolution," says Mr. J. E. Barnard, "is meant the power the objective has of separating and forming correct images of fine detail." Unless we see more of the intimate structure, the magnification in itself does not greatly avail. It does not help us to understand the thing better. Now there are two factors that determine this "resolving" power of the microscope. The first is what is called the "numerical aperture" of the lens, which means, in a general way, the number of divergent rays of light that the curvature of the lens will allow to impinge upon it. Lenses of high magnifying power are so small that they admit only a very small beam of light. Thus what is gained in magnifying power may be lost because of deficient illumination. A pretty device to increase the income of light in these high-power lenses was the "immersion lens," made of such a curvature that when the lens was focussed down into a drop of oil, or some other liquid, placed over the object on the slide, it received light from all sides. The drop into which the lens is focussed down or "immersed" greatly increases the illumination of a lens with high magnifying power. This method has enhanced the value of the microscope as an instrument that analyses structure, or, in other words, that discloses the intimate architecture of things. But the main point is that the "numerical aperture" of even the oil-immersion objective has at the present time reached its practical limit.

Yet there is a second factor, and that is the wave-length of the light-rays that impinge from the mirror and condenser on the object on the slide. But here again there is a limit, for, as Pro-

ONE OF THE FORE-LEGS OF THE MOLE CRICKET (GRYLLOTALPA)

Showing very strong digging blades, reminding one of the mole's claws. Small forceps are also seen, and an oval mark on the "knee" is a sense-organ sometimes regarded as an "ear." The Mole Cricket is nearly related to crickets and grasshoppers and is a very powerful burrower.

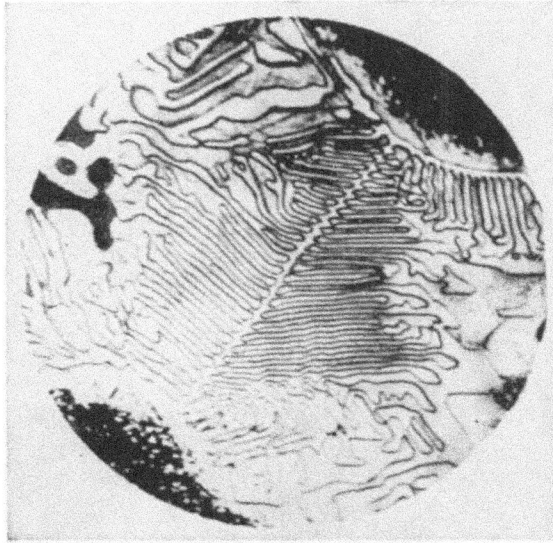

Reproduced by courtesy of Messrs. F. Davidson & Co.

A PIECE OF HIGH-SPEED STEEL SEEN UNDER THE
HIGH MAGNIFICATION OF 1,500 DIAMETERS

Note the marked lines of flow.

Photo: J. J. Ward.

SCALE FROM A GOLD-FISH

Showing the rings of growth by which the age of the fish can be
reckoned. Starting from a little speck, the scale has line upon line
added to it. Periods of rapid growth alternate with periods of slow
growth; or the growth in summer may be different in character from
the growth in winter; and thus alternate zones are established. Just
as we can tell the age of a tree by counting the summer and autumn
rings on the cut stem, so, after getting some secure data, we can
tell the age of the fish. It keep its diary on its scales.

fessor Bayliss tersely puts it: "Any object smaller than half the wave-length of the light by which it is illuminated cannot be seen in its true form and size owing to diffraction. Hereby is set a limit to microscopic observation." These are difficult matters, but the important point is that there are practical limits to what the microscope can do in the way of magnification and "resolution."

But Mr. J. E. Barnard has recently made an interesting step forward by using an illuminant such as a mercury vapour lamp, which is rich in blue and violet radiations. It may also be practicable to utilise invisible radiations in the ultra-violet, which would further increase the microscope's resolving power. As things are at present, the limit of useful magnification is somewhere about 800 diameters.

Beauty of Microscopic Structure

We cannot close this article without referring to a very different subject—namely, the extraordinary beauty of many microscopic objects. There are endless "beauty feasts" to be found in the architecture of the shells of diatoms, Foraminifera, and Radiolarians; in the structure of the outside of pollen-grains and butterflies' eggs; in the zoned internal structure of the stems of plants and the spines of sea-urchins; in the sculpturing of the scales on butterflies' wings and the multitudinous hexagons of their eyes; in the strange hairs on many a leaf and the elegant branching of zoophytes; in the intricate section of a rock and the variety of snow crystals. Of microscopic beauty there is no end.

BIBLIOGRAPHY

CARPENTER, *The Microscope* (1880).

DALLINGER, *The Microscope* (1891).

EALAND, *The Romance of the Microscope* (1921).

GUYER, *Animal Micrology* (1909).

LEE, BOLLES, *Microtomist's Vade-mecum* (7th ed., 1913).

SCALES, SHILLINGTON, *Practical Microscopy* (1909).

SPITTA, *Microscopy* (1909).

WRIGHT, ALMROTH, *Principles of Microscopy* (1906).

X

THE BODY-MACHINE AND ITS WORK

THE BODY-MACHINE AND ITS WORK

THE most perfect machine in the world is the body of man. The further we advance in our knowledge of it, the more we wonder at the ingenious mechanisms which are crowded into its structure. Almost every decade we discover new operations in it which have a profound influence over our life, yet are so subtle and unexpected that many generations of scientific men were entirely ignorant of them, and it may take still many generations to tell how they bring about their marvellous results. Here, as in other fields, the advance of science creates one mystery while it explains another. But that story of evolution which we studied in a preceding section sheds a clear light upon the body-machine both of animals and of man, and enables us to understand the high efficiency of most of its parts. The machine of the animal body is not only the most perfect in nature; it is immeasurably the oldest. For at least fifty million years—how much longer no man knows—the world-forces have been making the animal body, developing and improving the various organs and co-ordinating their functions. During all these tens of millions of years the machine has been subject to the fiercest stresses and trials, and the human body, as we know it, is the final and finished outcome. We wonder no longer. Time itself makes nothing; but if we grant a vast period of time to the real shaping forces of the universe, acting upon the most sensitive material in the universe, the perfection of the final stage becomes intelligible.

§ 1

Traces of the Past

We speak of the body as a machine, but it is hardly necessary to say that none of the most ingenious machines set up by modern science can for a moment compare with it. The body is a self-building machine; a self-stoking, self-regulating, self-repairing machine—the most marvellous and unique automatic mechanism in the universe. It differs from our ordinary machines, moreover, in this: when a part becomes superfluous or out of date, it will linger for ages, even for millions of years, in the structure, slowly changing and shrinking on its way to disappearance. It will be useful to begin our examination of the human body from this point of view, especially as some of the first things we notice about it are precisely shrinking structures of this kind.

Why have we hair on our bodies? We need not notice here the specially luxuriant growth on the head, or on the man's lips and chin. This has been artificially fostered or cultivated during the course of man's history. Men, in mating, chose women with rounded forms and smooth chins. Women chose men with strong muscular forms and, in our own branch of the race at least, hairy mouths and chins. We quite understand that in the course of tens of thousands of years this has evolved rich growths of hair in certain parts. But we have hair on our trunk and arms and legs; there are tiny pits in the skin, out of which hairs grow, all over the body except on the palms of our hands and the soles of our feet. Before birth the human body is, in fact, almost entirely covered with a fine coat of hair.

There is not a word to be said in favour of this part of our wonderful body-machine. It harbours dirt, microbes, and vermin, and sometimes favours skin-disease. As a coat it is ridiculously thin and ragged, and it has been superseded by clothing. Its plain meaning is in the story of life in the past which was told in an earlier part of this work. The hair is a dwindling vestige of the

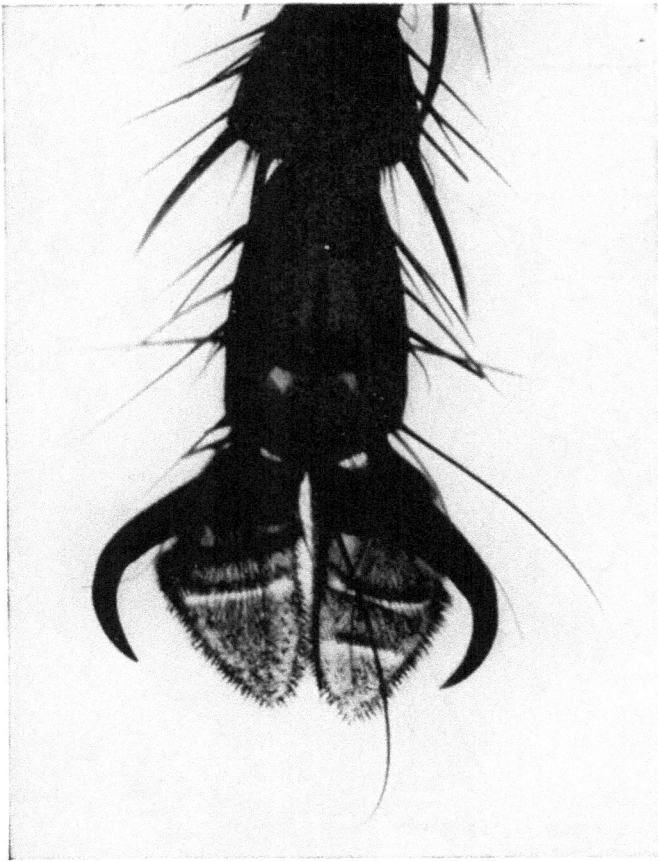

Photo: J. J. Ward.

FOOT OF A HOUSE-FLY

Showing two claws for gripping minute roughnesses and, between the claws, a double pad or cushion which is used in climbing up a smooth surface like a window-pane. The probability is that a moist secretion from the pad helps the adhesion. There is also very close apposition of the cushion to the smooth surface, and some say a partial vacuum is formed. But we are not quite sure how a fly runs up a window-pane—one of the most familiar sights in the world.

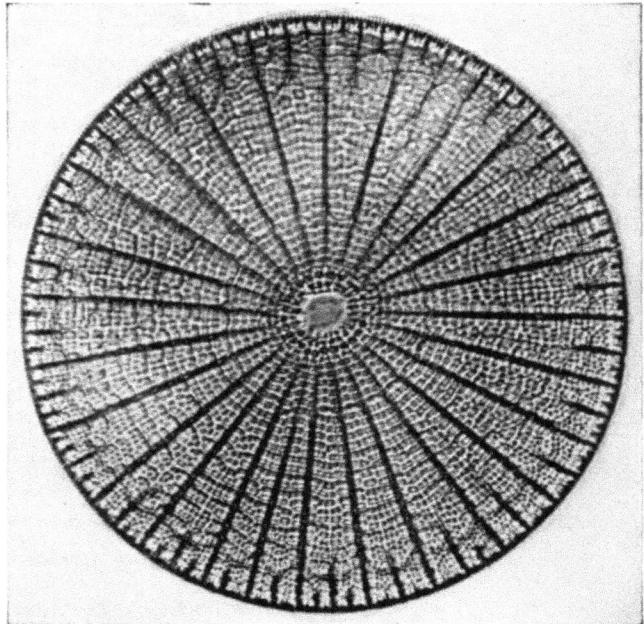

Photo: J. J. Ward.

THE BEAUTIFUL PATTERN ON THE MICROSCOPIC FLINTY SHELL OF A DIATOM (*Arachnoidiscus Ehrenbergi*), A KIND OF UNI-CELLULAR PLANT

Two hundred of these cells placed side by side would scarcely extend one inch. It is not known that there is any utility in these beautiful markings. They are expressions of rhythmic orderly growth.

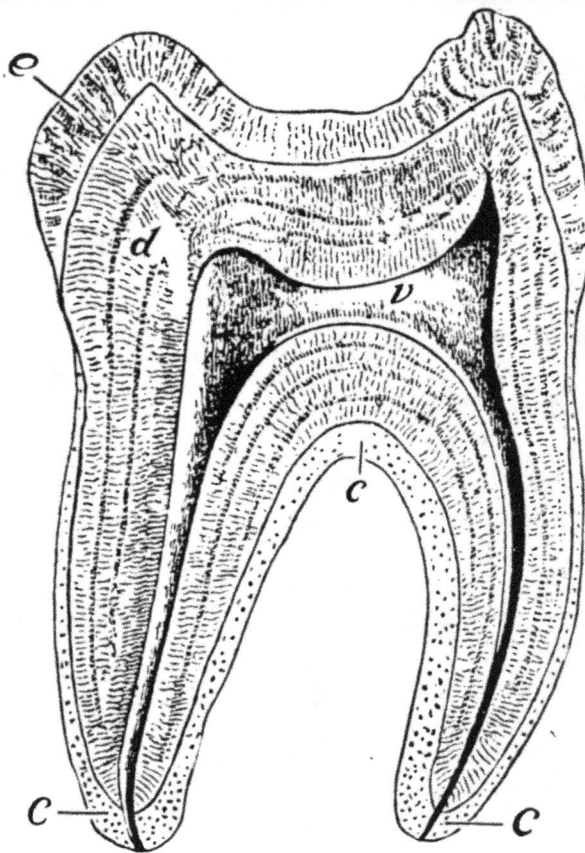

VERTICAL LONGITUDINAL SECTION (AFTER OWEN) OF A HUMAN MOLAR TOOTH

The lower part is embedded in a socket in the jaw and surrounded by the gum—the derma of the dense mucous membrane of the mouth. Two roots or "fangs" are shown, surrounded by a thin layer of bone called "cement" (c). Through the basal openings shown there enter blood-vessels and nerves. In the centre of the tooth is the vascular pulp-cavity (v), from which fine processes extend into the ivory or dentine (d). Over the crown there is the enamel (e), the hardest tissue in the body.

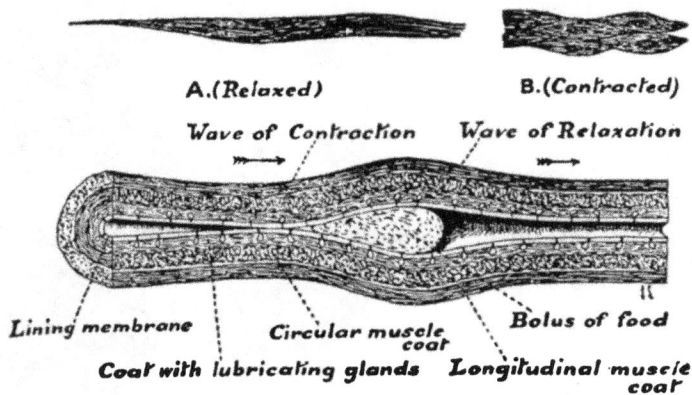

Reproduced by permission from Keith's "The Engines of the Human Body" (Williams & Norgate).

A PORTION OF THE ŒSOPHAGUS OR GULLET LAID OPEN TO SHOW A BOLUS OF FOOD PASSING DOWN

It will be seen that the muscular wall of the gullet relaxes in front of the bolus of food and contracts behind it.

A, a small fragment of the outer muscular coat magnified to show the individual spindle-shaped muscle fibres of which it is composed in a relaxed state. B, the same muscle fibres or spindles, greatly enlarged, showing the contracted state.

warm fur-coat which mammals developed to meet the conditions of an Ice Age. It is *vestigial,* not rudimentary, as is sometimes said.

The pieces of gristle or cartilage on the sides of the head which we call our "ears" are similar organs. They do not catch waves of sound, as many suppose, and guide them into the real ear inside the skull. They are too flat to do so. But if we compare them with the useful, pointed, movable ear of a horse, we see what they mean. They were once similar organs, but they have fallen out of use and are dwindling away. Underneath the skin we still have seven muscles attached to the shell of cartilage, from which it is obvious that the ear could once be moved in every direction to catch the waves of sound. Now only an individual here and there can use one or two of these muscles. The pinna or "ear-trumpet" is a surviving structure that tells us a little about the body's remote past.

There are very many similar muscles in the body to-day which merely tell us about a strange past. Some men can twitch their nostrils. Some can move their scalps. They do so by means of muscles which in most of us have gone completely out of use. Underneath the skin in very many parts of the body there are dwindling muscles of this kind.

In the inner corner of each eye we have a little pulpy mass which recalls to us even remoter ages of the body's past. It is of no use whatever in the body to-day. To understand it, one has to watch a parrot or an eagle in a cage, and notice how the bird flashes a white film (the "third eyelid") occasionally over its eyeball. Our superfluous bit of tissue is the shrunken remainder of this. We have in our eyelids a better apparatus for sweeping the dust off the eyeball, and the old membrane is disappearing. Man is not, of course, descended from birds, but almost all mammals have a well-developed third eyelid.

We know from fossil remains and from examining the bodies of living reptiles that in remote ages—somewhere about the era of the Coal Forests—there were animals with a third eye, in the top of the head.   We find this third or pineal eye in the heads of a few reptiles to-day, but the skin has grown over it, and it is degenerating.   In the birds and mammals it has sunk still deeper into the head, and degenerated further.   In man it has become a small body, about the size of a hazel-nut, rising from the middle of the brain.   We call it the "pineal body."   It is a mysterious little organ, and we will not say positively that it has no function. But, whether it has or no, we clearly trace it to the third eye of millions of years ago.

The "vermiform appendix" is a well-known vestigial organ. It is a little worm-like tube, about four inches long, arising as a blind alley at the junction of the small and the large intestine; and as a source of disease (appendicitis) and danger it is notorious.   Some have tried to find that it has a use in the body, but the plain fact is that it has been removed from hundreds of thousands of men in modern times, and no harm has ensued in any single case.   It is the vanishing remainder of a large, useful chamber in which early vegetarian mammals let myriads of bacteria break up their coarse food.

In short, expert investigators of the human body have found 107 organs, or parts of organs, that have to be understood as more or less vestigial.   We have the vestige of a tail in the block of bone at the lower end of our backbone; and sometimes children are born with distinct and movable—though very short—tails. We have bones, muscles, and glands in many parts that are now the almost or quite useless relics of a remote past.   Evolution beautifully explains them.   The body remains a wonderful mechanism because these were once useful and most cunningly contrived structures.

§ 2

The Ivory Gates of the Body

Turning now to the body-machine in its active life, we shall find it most interesting to follow the progress of food until it is built into the frame. We can, if we like, do this literally to-day. We can mix an opaque powder, e.g. of bismuth, with a mouthful of easily digested food (such as meal), and then by means of X-rays we can photograph its progress along the alimentary canal until what remains of it is dumped in the waste-chamber. This is useful for some purposes, but in the main we rely on the minute studies of the anatomist and the experiments of the physiologist for our knowledge of that part of man's body that is concerned with the utilisation of the food.

The receiving office, so to say, or the mouth, is itself so deeply interesting and full of ingenious contrivances that a whole section of this work might be devoted to it. Above the mouth are the sentinels, the eyes and nose, which we shall consider later. Then, on the upper surface of the tongue, there are myriads of sensitive little organs, taste-buds, ready to apply a final test to the food. The juices of the food penetrate the thin skin which covers and protects them, and probably have a chemical action on the little nerve-endings in them. If the message which they automatically dispatch to the brain is "O.K." other nerves set in motion the muscles of the lower jaw, and the grinding of the food begins.

In modern times a volume could be written on the teeth alone, and it would be a remarkable story. Teeth began ages ago in fishes of the shark and dog-fish type. Their rough scale-covered skin was drawn in to form a lining of the mouth, and long practice in crunching shell-fish, and so on, led to the evolution of the hard scales on the skin into teeth. In the course of time the teeth on the ridges of the jaw were particularly developed, and the others disappeared. But we must not suppose from this brief hint of their origin that teeth are simple things. Each tooth is a remarkable structure. Coats of dentine and enamel are built round a pulpy cavity into which nerves and blood-vessels run, and roots, coated with cement, fix the tooth in its socket. It is strange how few people wonder why our jaws do not ache and jar from the crunching of a hard crust. In mammals there is a provision which obviates this: the teeth do not fit tightly in the sockets. They are "packed" in with a material that lessens the shock of the daily grind.

A special regiment of the cells which make up the body is told off to see to this business of the teeth, and our wonder increases when we see these strange microscopic, unconscious "bone-builders." They have, in the baby, to select, atom by atom, the cement and dentine and enamel which (in other forms, of course) somehow exist in the food of the blood-stream. They have to build the stuff into structures of which our artificial teeth are very clumsy imitations. They have to do this at the right time—to wait until suckling is over and eating begins. Then they have an even more difficult problem to face. The jaws of the adult will be much larger than the jaws of the young, and, naturally, it is not possible to alter these solid structures of enamel and dentine. So the bone-builders absorb the greater part of the first set of teeth, the "milk-teeth," and meantime prepare a new set underneath them. The cast tooth which your child shows you is, as a rule, a mere shell. The microscopic bone-builders have re-absorbed the material.

Yet the teeth, with all their wonders, may be among the doomed structures of the body. Some authorities believe that they will gradually drop into the class of vestigial organs, like the hair, though they will give a vast amount of trouble as they grow weaker. Their purpose is to break up the food into smaller particles, and in modern civilisation this is done in the preparation and cooking of the food. So the teeth are already going. In ancient skulls and among savage peoples the teeth are often much worn away by sheer hard work; in conditions of civilisation much-worn teeth are rare. The jaws have not the hard work they once had; they therefore get a smaller supply of blood, and the teeth grow weak and less numerous. Normally we have thirty-two teeth, but in many people the so-called "wisdom teeth" (which might very well be called unwisdom teeth, as they are

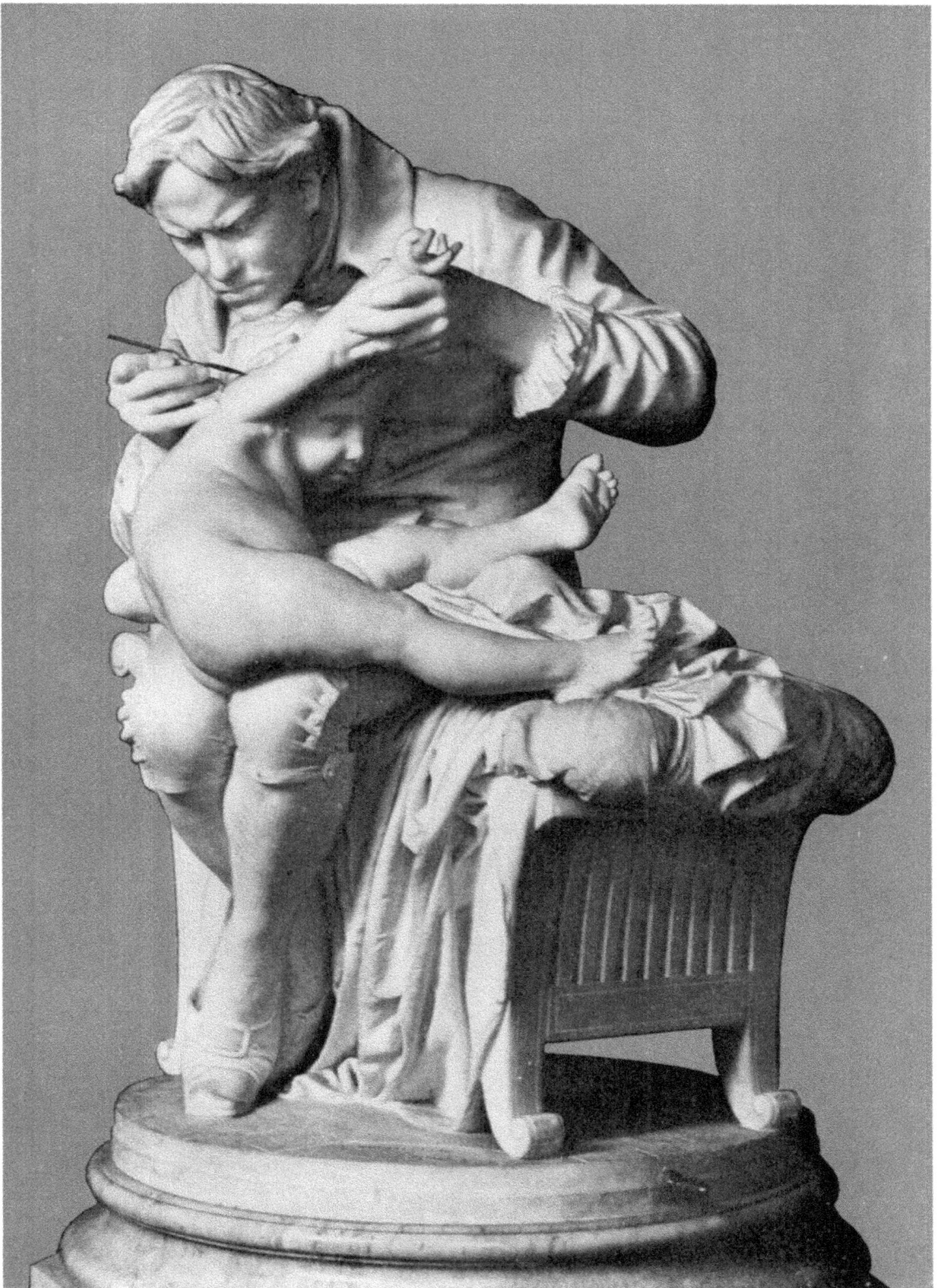

DR. JENNER (1749–1823) INOCULATING HIS OWN SON: THE STATUE BY J. MONTEVERDE AT GENOA

He proved about 1796 by a long series of observations and experiments that vaccination with cowpox protects the body from smallpox, either completely or to the extent of diminishing the severity of an attack.

superfluous) do not develop. On the other hand, a few people have thirty-six teeth, and when we turn to the monkeys we find that some of them have this number. We are, in fact, slowly shedding our teeth, in fours, along the corridors of time, and it may be that in the distant future man will be toothless. Perhaps, however, a fresh enthusiasm for physical vigour will save the race from any such degeneracy.

But our "grinding-stones" are only part of the mechanism of the mouth. As soon as the grinding begins three pairs of glands pour saliva into the food. Here again we have an automatic nervous machinery of a remarkable kind, for, as everybody knows, the mere sight of food may set the glands at work, or "make your mouth water." In the glands themselves the microscopic cells which make the saliva are such remarkable chemical machines that we cannot yet understand them. They not only help to make a soft pulp of the food—the saliva is 99 per cent water—but they pour into it certain chemicals which begin the digestion of starchy food, converting it into a kind of sugar. That is one of the reasons for thorough "mastication." One must not suppose that it does not matter, since the food is so short a time in the mouth that there cannot be much chemical action on it. The chemical action of the saliva *goes on,* for half an hour or more, while the food is in the stomach. A great deal of illness is due to neglecting to mix plenty of saliva with our bread, etc., before it leaves the mouth.

§ 3

The Process of Digestion

When teeth and salivary glands have done their work, and the taste-buds on the tongue have had their moment of satisfaction, the mouthful of pulp is swallowed. "Swallowing seems such an easy and automatic act that we are quite unaware of the elaborate system of signals, side-shunts, and level-crossings which have to be manipulated to permit the busy traffic of the pharynx to pass unchecked." Food does not "go down," as children think. The whole mouth changes. Certain sensitive spots at the back of the mouth—electrical press-buttons, we may call them—give the signal that the food is ready. The powerful muscles which closed the back of the mouth while we are masticating, relax. The lower jaw is pressed against the upper. The soft palate forms an inclined plane. Other muscles close the airways to the nose and the great airways to the lungs. The whole complex machine acts together and pumps the food into the first part of the alimentary tube, the pharynx. Only very rarely does a little food "go the wrong way"—into the air-passage—and then another set of muscles automatically blow (or cough) it out. The mouth is a rather complicated cavity. How many communications it has—with the nasal passage by the posterior nostrils, with the ear-passages by the Eustachian tubes, with the pharynx, and with the windpipe through the guarded glottis. It is obviously very important that the mouth be kept in good order.

The food now enters upon a very long journey. Most people are surprised to hear that the alimentary tube is yards long in a man or woman of medium height, about twenty-eight feet long from the mouth to the vent. But few people reflect—there would be far less misery and discomfort in the world if they did—what digestion really means. Our food has to be broken up, physically and chemically, and even then it cannot pass into the blood. From the pulpy mass myriads of tiny organs have to select the matter which the body needs, and let the remainder pass away. So the food has to pass slowly through twenty feet of a long tube, the small bowel or small intestine, to give these "selectors" the chance of taking the nourishment from it.

However, let us begin at the beginning. The food passes to the stomach down a one-foot tube (in a man of medium height), the upper part of which is known as the pharynx, and the lower part as the gullet or œsophagus. As we said, it does not slide down.

Sir Arthur Keith describes very graphically the transport of a bolus of food towards the stomach. "The instant that a bolus has been pushed through the doorway leading from the pharynx, and that doorway has closed, we see a ring of contraction form behind the bolus and commence to creep slowly downwards, forcing the bolus in front of it. The bolus, on entering the œsophagus, has touched a 'button,' and the ring of contraction is the result. As the bolus is driven forwards it comes in contact with a succession of such buttons, with the result that it is kept moving onwards. Not only so; a ring of relaxation precedes the bolus and eases the passages." [1]

The earliest stomach in the animal world was merely a straight tube through which the food passed, but in the course of evolution it has grown larger and larger at this spot until (in man) we have a large storage chamber in which the food will be churned up and mixed with acid and ferment during several hours. The stomach goes up close to the heart on the left side, but its large upper part is less concerned with digestion. The muscular movements which push the food about, so as to expose every part of it to the digestive juices, begin about half-way down the stomach and travel, in waves, towards the bottom. There are three coats of muscle, and they are at work all day long mixing the pulpy contents. The stomach of a healthy and sensible man will get through its work in about four hours, and, if he postpones his next meal, it will, like Oliver Twist, call for more. It develops a peculiar writhing motion in its muscles, and this is telegraphed to the brain by the nerves. You feel "hungry."

The inner wall of the stomach is richly supplied with blood, and is lined by the myriads of minute glands which produce the "gastric juice." As soon as you sit at table, the sight and smell of the roast mutton send their messages to a certain nerve-centre, and from this a silent message of stimulation goes to the glands.

[1] Sir Arthur Keith, *The Engineers of the Human Body.*

When the food touches the taste-buds on the tongue, the messages multiply. The blood gathers in the wall of the stomach, and the little tubes use its nourishing solution in order to make the digestive juice which they pour out upon the food. With a large part of our food the stomach does not deal. Digestion only begins in it. Sugar, starches, and fats are passed on to the next department. It is chiefly the nitrogenous or protein food—flesh, fish, eggs, etc.—that is here broken up still further and prepared for absorption. The stomach itself absorbs only a little of the food. A glass of wine—as the head of an inexperienced maiden may tell her—is absorbed into the blood very quickly, and part of the digested meat is passed into the blood-system through the stomach; but the main part of the food passes into the twenty-foot laboratory of the narrow "small intestine," to be further digested and absorbed.

It is amazing how few people have even a rudimentary knowledge of this fundamental part of their being. Stomach and bowels are hopelessly confused, and our poor organs, magnificent as they are, are treated with inconceivable crudeness even by educated people. It is easy for everybody now to obtain a list giving the exact digestibility and nutritive value of different kinds of foods. It should be known to everybody that the work of these myriads of minute chemical laboratories in the stomach has drawn the blood from the brain for a time, and it is unnatural and unhealthy to attempt brainwork during or after a meal. Half the physical misery of life could be cured by a little knowledge and restraint.

§ 4

A Remarkable Mechanism

A brilliant physiologist, Professor Metchnikoff, startled us some years ago by stating—almost snorting—that the human alimentary system is a miserable out-of-date machinery that ought to be scrapped. Even the stomach, he said, was super-

THE SCHOOL OF ANATOMY, BY REMBRANDT

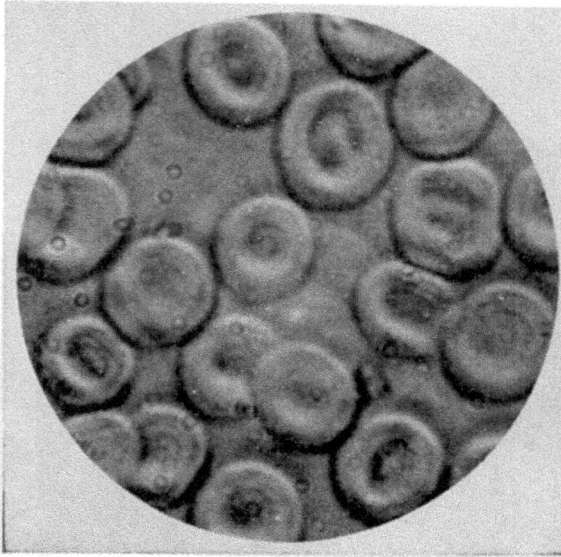

Reproduced by courtesy of Messrs. F. Davidson & Co.

RED BLOOD CORPUSCLES OF MAN

Each cell or corpuscle is a circular disk, on an average about $\frac{1}{3200}$ of an inch in diameter, and about one-fourth of that in thickness. More than a million will lie on a square inch. The broad faces are not flat, but slightly concave; so the corpuscles are thinner in the middle than at the margin. When seen edge-on they look like rods. Their colour is faint yellowish-red, due to the pigment hæmoglobin, which has a great affinity for oxygen. The mammalian red blood corpuscle does not show any nucleus except in the early stages of development. The white blood corpuscles are larger, nucleated, and irregular. The red blood corpuscles are mostly made in the marrow of the bones and mostly destroyed in the liver and spleen.

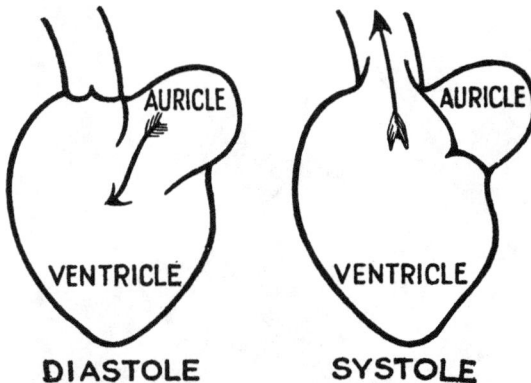

DIAGRAM ILLUSTRATING THE VALVES OF THE HEART

During the period of contraction (systole) of either of the ventricles the valves which guard the opening of the big artery are open, whilst the valves at the gateway between auricle and ventricle are closed, thus preventing blood from being forced back into the auricle. During the period of relaxation (diastole) of the ventricle, the valves of the artery are closed and those between auricle and ventricle are open, allowing blood to flow freely from auricle to ventricle, but preventing back-flow from the artery.

Both auricles contract at the same time. Then both ventricles contract simultaneously. Then there is a short pause. Each complete cycle, including the pause, is a "beat"; and in a healthy adult the heart beats about seventy-two times in a minute.

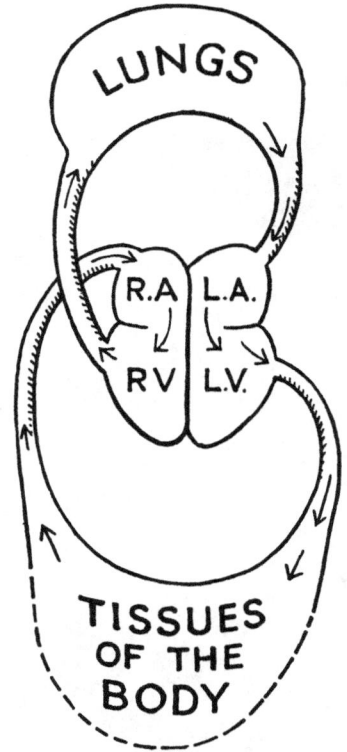

DIAGRAM OF THE CIRCULATION OF THE BLOOD THROUGH THE BODY

Seventy-two times a minute the heart beats; that is to say, its muscular walls contract. Each half of the heart consists of two chambers: the auricle, which receives blood from the big veins, and the ventricle, which receives blood from the auricle and pumps it into the big arteries. Between the right and left sides of the heart there is no direct communication. Aerated blood is collected from the capillaries of the lungs and conveyed by the pulmonary veins to the left auricle, whence it passes to the left ventricle. Thence it is pumped into the great artery, the aorta, whose branches distribute it to every part of the body. Having given up much of its oxygen and nourishment to the tissues needing them, the blood is collected and conveyed by veins to the right auricle. From there it passes to the right ventricle, which forces it through the pulmonary arteries to the lungs where it is aerated and again travels to the left side of the heart, ready to be again circulated. It will be noted that arteries are blood-vessels leaving the heart, while veins are blood-vessels returning to the heart. The arteries carry pure blood, except in the case of the pulmonary arteries to the lungs. The veins carry impure blood, except in the case of the pulmonary veins from the lungs.

fluous. Very few men of science agree with him, but when we understand what he was driving at, we see that he had hold of a very important idea. Professor Metchnikoff meant that we can "digest" our food in advance by means of chemicals, and get rid of all the great length of intestine which is so liable to disease. At present we take in great masses of superfluous stuff, but there will some day be extraordinary changes in man's diet. In the meantime the organs are there and we have to feed accordingly; but it does not follow that this will continue.

At the base of the stomach, where the alimentary canal again becomes a tube—the beginning of the small (or narrow) intestine—there is a very powerful ring-muscle, which guards the exit. Nature has provided for healthy living. The unprepared food, if we live well, cannot run into the bowel. Mere contact of food against this muscle only draws it tighter. There has to be, apparently, some sort of chemical action by food which is quite ready, then the muscle relaxes, and a spray of the pulp shoots into the bowel. We can see the occasional gush when we are following the digestion by means of X-rays.

In the first section of the narrow intestine, the duodenum, very important work goes on. Here we find a remarkable sort of mechanism which has only been discovered by recent science. The automatic machines of the body generally work by nervous action. The sight of roast beef is "wired" to the brain by the eyes, and a reflex (or reflected) nervous action sets the salivary glands and the stomach glands at work. This is a sort of automatic telegraphic system. But there is also in the body something like a postal service; it is one of the most wonderful discoveries of modern physiology. When, for instance, the acid food (or chyme) from the stomach touches the walls of the lower bowel, the glands in this wall form a certain chemical (called *secretin*) and pour it into the blood. The blood takes it rapidly all over the body, *but there is only one organ (or perhaps two) waiting for it.* In this case the pancreas (or sweetbread) receives the chemical message, and sets more vigorously to work pour-ing an increased supply of digestive juice into the intestine. These marvellous messages which are, as it were, posted in the blood are called "hormones," and, as we shall see later on, we find them again and again accounting for very remarkable results in the body.

The liver and the pancreas are really outgrowths from the alimentary tube; sections which have become detached, special departments, with ducts leading to the bowel. Each pours about a pint of fluid daily into the bowel, to assist the work of digestion. The liver has very special work to do, and we shall notice it presently. Here we need only say that the bile which it pours into the bowel—sometimes so abundantly that some forces its way into the stomach, and you get "bilious"—is not a digestive juice, but it helps to prepare the fats in the food. The pancreatic juice, on the other hand, is a real digestive juice, and the starches and sugars and fats, as well as the nitrogenous foods, are now dissolved and made into emulsion, and prepared, in short, for absorption. The juices of the pancreas and of the intestine include powerful ferments, or *enzymes*. These are substances which cause chemical changes by their mere presence, *without being used up in bringing about these changes*. One of them deals with the starches and sugars in the food, another with the fats, another with the proteins, like white of egg.

The chyme—the pulped and semi-digested food—moves slowly along the bowel. In the bowel wall are muscles which contract, and drive the contents slowly onward at about one inch a second. But the interior wall of the tube is now lined, not only with glands, but with little outstanding fingers, like (on a microscopic scale) the pile on fine velvet. The surface is, moreover, puckered into folds, to give a larger area, and myriads of the tiny fingers or *villi* dip into the chyme and absorb the nourishing matter. Altogether about sixteen square feet of absorbing surface are given in the small intestine, and it is there that most of our nourishment is taken into the blood or the lymph. The rest passes on into the "large intestine," a very much wider tube at its lower extremity.

It is just at this junction-point of small and large intestine that the vermiform appendix is given off from the tube. Its opening into the bowel is very narrow, and sometimes bits of hard food, such as fruit-seeds, get into it and set up inflammation. Now in certain vegetarian animals like the rabbit the appendix is at the end of a large and useful blind alley or cæcum. A good deal of vegetable matter is wrapped up in cell-walls of cellulose, and the digestive juices we have described have very little power to deal with cellulose. It has to be broken up by bacteria, in such a chamber as the rabbit's blind gut or cæcum. The human vermiform appendix is a remnant of the large magazine in which some coarse-feeding ancestor of man had the cellulose of his food dealt with by bacteria.

Most people are inclined to shudder when they hear of bacteria in their bodies, but it is only certain kinds of bacteria that pour poisons into the blood and cause disease. Each one of us houses trillions of friendly bacteria in the large intestine. They do us no harm, however, but break up the cellulose (the husks of grain, etc.) and multiply prodigiously in the fetid, fermenting mass in the intestine. Some physiologists think that we should be better without the large intestine, but, while there is little food absorbed from it, much of the water we take up is absorbed there. In any case, there the structure is, and the wise man takes plenty of cereals, fruit, and green vegetables in his food, to keep it in a state of healthy activity.

§ 5

The Vital Fluid

Let us return to the nutritive material which has been taken up into the system. The tiny organs on the bowel wall which absorb it pass most of it directly into the blood-vessels which they contain. It becomes part and parcel of the blood, and will now, after passing through the liver, be pumped round the body *for the various organs to select from it the nourishment they need.*

The blood is not the simple matter which it seemed to our ancestors; indeed, it has proved of late years to be full of subtle interest. If you prick your finger with a needle, and put a small drop of blood under the microscope, you see myriads of little disks, many of them in rows like columns of pennies, in a watery or yellowish fluid. The fluid—the serum or plasma—is the liquid food of the body and the medium for conveying away the soluble waste matter. The red disks or "corpuscles" are the bodies that convey oxygen from the lungs to the tissues. There are about five millions of them in every cubic millimetre of the blood of a healthy man—a woman's allowance is half a million less—and it is these which give the blood its red colour. We have about twenty-five thousand trillions of them in all.

Thus the microscope discovered a new and unexpected complexity in the blood, and further research has shown that it is the iron-containing pigment in these red corpuscles which is chiefly concerned in the carriage of oxygen. There is very little iron in the blood, and it is absurd to think that we increase its quantity (once it is normal) by eating things "with iron in them," as people say. But what iron there is may be called a precious metal in the human body, and the red corpuscles have it in a form that still more or less baffles the chemist. There are believed to be something like two thousand atoms to the molecule in the red pigment (hæmoglobin) of the corpuscle! These disks, or corpuscles, are formed, mostly, in the red marrow of the bones, and, after serving for a few weeks, they are broken up in the liver or the spleen.

Warfare in the Body

But this is only the beginning of the interest of the blood. In the drop which we observe under the microscope there are, as

Photo: *Rischgitz Collection.*

HARVEY DEMONSTRATING THE CIRCULATION OF THE BLOOD TO CHARLES I.

(From a Painting after Robert Hannah.)

William Harvey, 1578–1657, a graduate of Cambridge, who studied under Fabricius, the great anatomist of Padua. He demonstrated about 1616 that the blood passes through the body in a "kind of circle"; but he did not know of the capillaries which connect the arteries and the veins. He insisted on experiment as a basis of knowledge, and he made important observations on the development of the chick.

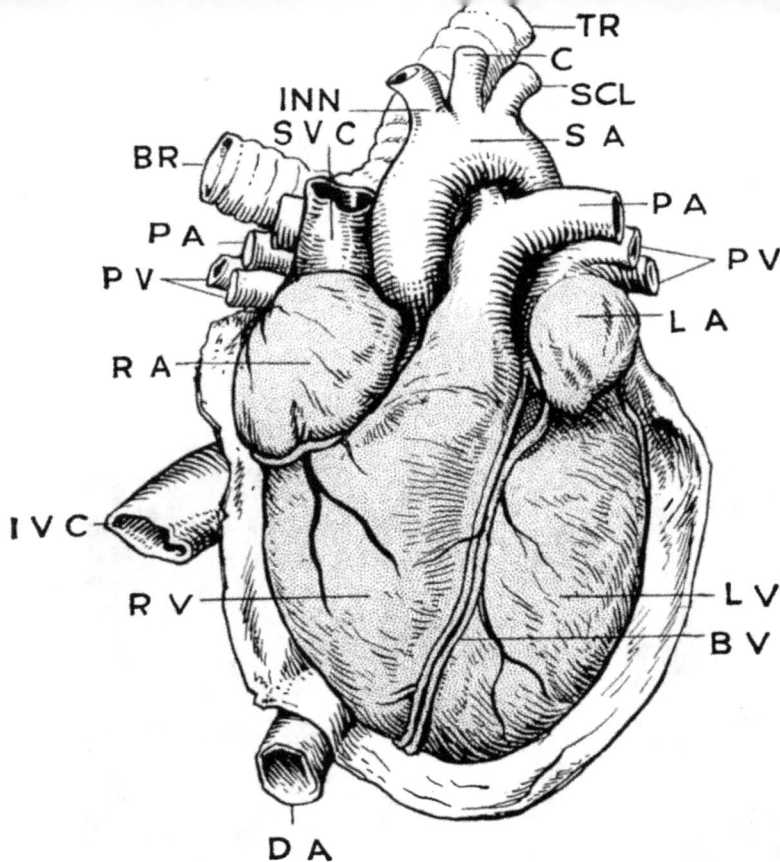

THE HUMAN HEART

There are four chambers, two auricles or receiving chambers and two ventricles or driving chambers. The right auricle (RA) receives the impure blood from the body by two superior venæ cavæ (SVC) and one inferior vena cava (IVC).

The blood is passed into the right ventricle (RV), whence it is driven to the lungs through the pulmonary arteries (PA). From the lungs the purified blood returns by the pulmonary veins (PV) to the left auricle (LA).

From the left auricle it passes to the left ventricle (LV), whence it is driven up the systemic arch (SA) to the body. The systemic arch first gives off a right innominate artery (INN, dividing into right subclavian and right carotid, to arm and head respectively). It then gives off a left carotid (C) and a left subclavian (SCL), to head and arm respectively. It is continued dorsally backwards to form the dorsal aorta (DA), the great artery distributing pure blood to the whole posterior body.

TR is the windpipe or trachea; BR, a bronchial tube carrying air to the lungs; BV, a blood-vessel on the wall of the heart itself.

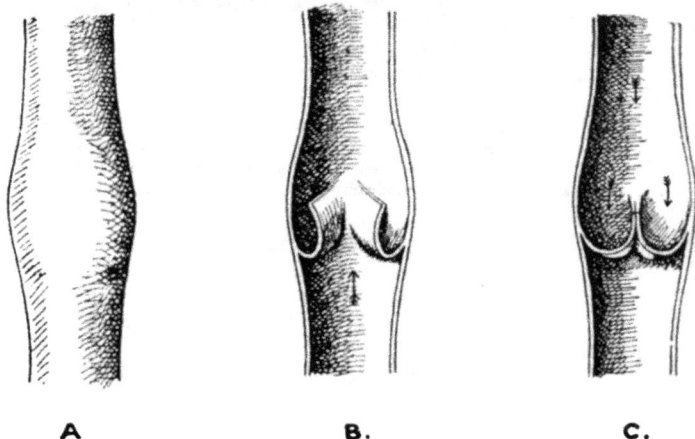

A.   B.   C.

Reproduced by permission from Keith's " The Engines of the Human Body" (Williams & Norgate).

A, a swelling on a vein, indicating the presence of a valve within it. B, the vein laid open, showing the valves partly open: blood flowing in the direction of the arrow will have free passage between the valves. C, showing the valves shut: blood forced backwards in the direction of the arrows will find the valves

we said, myriads of red corpuscles in a yellowish fluid. It was
found out some years ago that the serum is quite congenial to
its own corpuscles, but that if we mix into it a little of the blood
of some animal of a different kind, the serum of the first animal
destroys the red cells of the second animal. Thousands of experiments were made, and it was found that the degree of action of
one kind of blood upon another depended on the nearness or
remoteness of relationship of the two animals. If they were
nearly related, there was no destructive action. Naturally, the
opportunity was soon sought to apply this new test of "man's
place in nature," and it was found that his blood mingled amiably
with that of the anthropoid apes!

There is a third element in our drop of blood under the
microscope, and this is the most interesting of all. Here and
there in it, though hundreds of times less numerous than the red
disks, there are what are called "white corpuscles," microscopic
colourless roundish specks. When we study these specks closely,
we find that they behave just like the very primitive microscopic
animal known as the *Amœba*. They push out parts of their substance, and glide along. If there are bacteria in the blood, one
of these corpuscles may be seen making its way to one of the
intruders and slowly folding its substance round it. After the
microbe is engulfed, digestion soon follows.

In other words, there is in the blood, besides the army of
oxygen-carriers, a great army of defenders against bacteria. Let
a tissue be injured somewhere, and the injurious bacteria find a
footing and begin to multiply at an appalling rate. We are
threatened with disease, if not death. The bacteria may destroy
the tissue, or pour poison into the blood. But the "white knights"
now gather from all parts to defend the body. They are brought,
of course, by the flow of the blood, but they seem to have some
sort of chemical sense for bacteria, and they crowd in the particular tissue which is threatened. A great struggle ensues, and
the patient's temperature rises to "battle heat." If the white

corpuscles succeed in devouring the microbes before they multiply to a dangerous extent, we are saved. But bacteria multiply at a terrible speed, and sometimes they beat the corpuscles and we pass into a perhaps dangerous illness.

Biologists had hardly ceased to wonder at this new romance of the blood when others were discovered. Bacteria produce a poison (or *toxin*) with which they taint the blood. But it was found that the blood produces an "anti-toxin," a chemical for neutralising the toxin; and after years of experiment the anti-toxin was prepared in the laboratory and injected into the blood. It also became possible to help the white corpuscles in the fray, or spur them on to it, so to speak, by preparing a sort of sauce —an *opsonin,* the man of science calls it—from dead bacteria and injecting it into the blood. The opsonin makes the living bacteria more attractive or palatable to the corpuscles, and our "brave defenders" go to work more vigorously. Sir Arthur Keith, in *The Engines of the Human Body,* refers to "the immense and movable armies of microscopic corpuscles which can be mobilised for police or sanitary duties. They swarm in the blood stream as it circulates round the body . . . it is extremely probable that one variety of them, if not more, are really errand-boys on their way to deliver messages or parcels, and that the gland masses which are built up in and around lymph channels serve both as nurseries for the upbringing of such messengers, and also as offices from which they are dispatched on their errands."

§ 6

The Heart

It is clear that the many-sided value of the blood depends upon its regular coursing through the whole body, and we have now to see how this is accomplished.

Until three centuries ago there was not a man in any civilisation who knew anything about this "circulation of the blood."

The most learned physicians had the weirdest ideas about the function of the heart and the flow of the blood. Nowadays the essential facts are familiar. The heart, which one feels beating about the lower part of the breast-bone (though drawn a little to the left), is the central pumping-station. From it goes a great tube, or artery, which branches out—much as the trunk of a tree divides into branches, and finally into twigs—until its finest ramifications have carried the blood into the remotest tissues of the body, even into the teeth and bones. There the little twigs turn back, as it were, and become veins, and the veins from all parts join each other and at last bring the blood back to the heart.

In a sense it is as if the fresh-water circulation and the sewage circulation of a great city were managed from the same pumping-station. One set of pipes would convey water to every tap; another set would bring back the foul water to the pump. The difference is that in the animal body the two sets of pipes join on to each other and form a continuous system. But, obviously, fresh and foul blood must not mix, and this has been secured by the evolution of a heart with the two halves completely separated from each other. We can trace the evolution of the heart by studying it in various types of lower vertebrates. In most reptiles the two halves are still imperfectly separated, and "mixed" blood (pure and impure, fresh and foul) goes to the greater part of the body. In the mammals and birds the separation is complete.

The heart is a thick muscular pouch—with walls about half an inch at their thickest part in man—which has to drive the blood to the tissues on the one hand, and to the lungs for purification on the other. That is why it has separate halves. Each half, moreover, has a little chamber for receiving the blood (an auricle) and a larger chamber for pumping it (a ventricle), and valves are cunningly contrived at each opening so that the blood can flow only in one direction when the pump works.

So remarkable is the mechanism of the heart, that we do

not yet know what regulates its "beat." There seems to be some mechanism in the heart itself for regulating it. Seventy-two times in every minute, in a healthy and resting man, the chambers draw their walls together and pump out the blood. There are tens of thousands of muscular fibres built so wonderfully into the walls that the chambers can close in from every side, like a man closing his fist, and give the blood a start that will carry it all round the body and back to the starting-point.

It is, of course, a mistake to say that the heart never rests. It rests, and recovers, between each beat. But its function is remarkable. As we said, it beats seventy-two to the minute when a man is resting. But let there be some sudden call for action, and almost before you get from your chair, the great pump beats faster, as if it knew that the distant muscles and brain had now work to do and must have more blood. When we are sitting still, it throws five pints of blood (a little more than a third of all the blood in the body) into the arteries every minute. During a quick walk the heart pumps seventeen pints a minute; and the man who runs upstairs is asking his heart to pump thirty-seven pints a minute! During even less violent exercise than this, all the blood in the body (about fourteen pints) passes twice through the left ventricle of the heart and all round the body in a single minute.

From the left ventricle, the chief pump, the blood passes into a thick broad tube called the aorta. The elastic walls of this tube expand as the blood rushes in, and then slowly close again, driving the blood onward. In this way, and by the general resistance of the tubes (the arteries), the jerky discharge from the heart is converted into a steady flow after a time. The arteries branch out in every direction, and as they approach the tissues they have to feed they break into myriads of very fine tubelets, often not more than 1-3000 of an inch in thickness. The wall of the blood-vessel has now to be so thin that the nourishing matter in the blood can flow through it to the tissues, and the

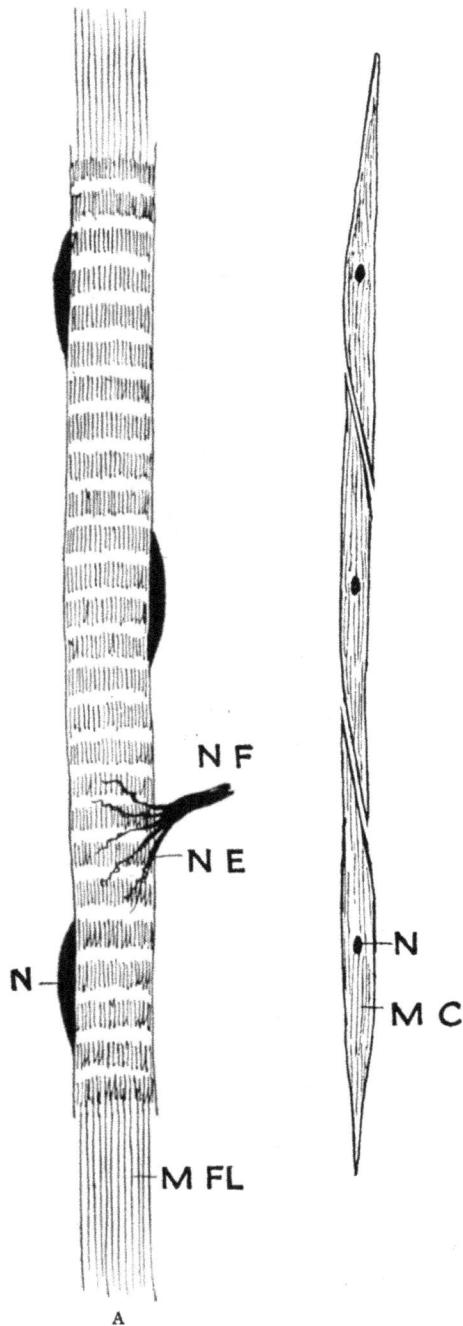

A

A. A striped or striated muscle-fibre, quickly contracting, showing alternate dark and light bands. It is built up of very delicate fibrils (M FL). It is stimulated by a nerve-fibre (N F), which divides into an end-plate (N E) on the contractile substance. A striped muscle-fibre is due to the great elongation of a cell, with multiplication of nuclei (N), or to a fusion of elongated cells.

B. Three smooth or unstriped muscle-cells (M C), elongated spindles, dovetailed into one another, each with a nucleus (N). There may be longitudinal fibrillation. Smooth muscle-cells are slowly contracting. They occur in such situations as the wall of the food-canal, the wall of the bladder, the wall of the arteries; and abundantly in sluggish animals, such as sea-squirts.

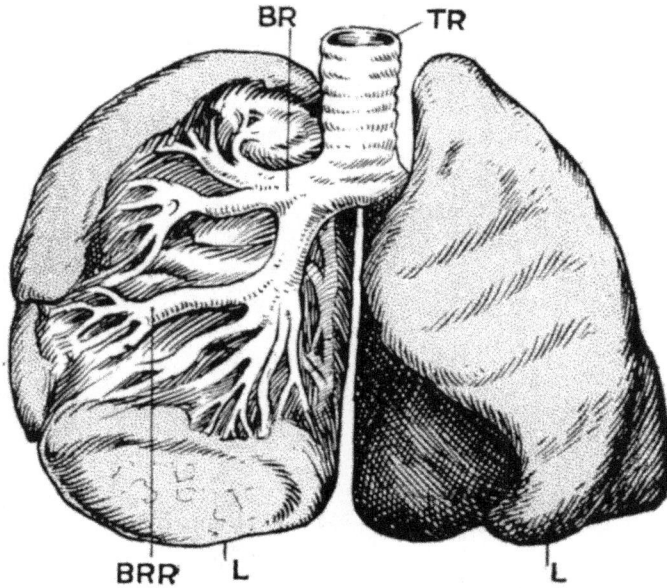

THE WINDPIPE AND LUNGS OF MAN. (From a specimen)

TR, the trachea or windpipe, supported by gristly rings. It divides into two bronchi (BR) entering the lungs (L). There they break up into finer and finer bronchial tubes or bronchioles (BRR), ending in little dilatations which are divided into chambers called "air-cells." On the walls of these the interchange of gases takes place. To the left side, as diagrams go, the lung is seen intact; to the right, partly dissected.

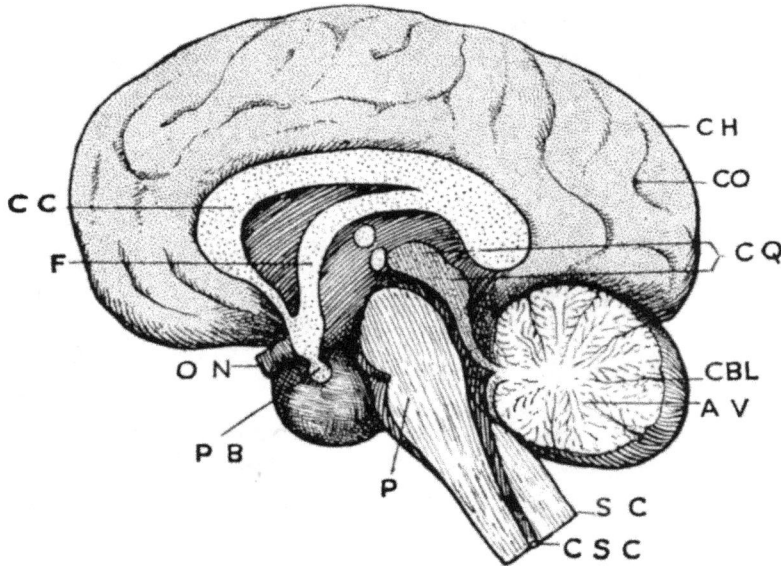

SECTION OF A HUMAN BRAIN

Nerves entering the brain. F is a longitudinal bridge of fibres, called the fornix. It makes the roof of the optic thalami region, or third ventricle. Behind the fornix are seen two transverse commissures cut across. CQ, the corpora quadrigemina or optic lobes. CH, the forebrain or cerebral hemispheres, showing internal indications of convolutions (CO).

CBL, the cerebellum, with an internal pattern (AV) called the arbor vitæ. This is due to the folding of the nervous tissue into a number of lamellæ, which give off secondary plaits.

SC, the spinal cord. CSC, the cerebro-spinal canal continued down the centre of the spinal cord.

P, the pons Varolii, a bridge forming a sort of transverse floor to the cerebellum. Behind P is the "bulb" or medulla oblongata.

PB, the pituitary body, a nervous and glandular body growing down from the floor of the optic thalami region, or third ventricle. ON, the optic nerve.

waste matter from the tissues can get back into the blood. Even this is a far more complicated matter than is generally supposed. The cells in each tissue of the body must somehow select their own food and oxygen, and even the union of oxygen with carbon in the working muscle does not take place as we find it in ordinary combustion.

A Wonderful Apparatus

At the point where the artery subdivides into the finest tubelets (the capillaries), there is a wonderful apparatus, a sort of "stopcock," for regulating the supply of blood. Muscular fibres are coiled round the artery, and, as the artery enlarges or contracts, the supply of blood to that particular tissue is increased or lessened. When you sit down to a meal, for instance, the stopcocks are opened full to your digestive organs and partially closed against your muscles and brain. When you stand up and move about the room various muscles have to work, and the cocks are duly turned on to them. When your muscles need all the blood they can get, your brain and digestive organs get less. When you stand erect for some time, the regulative system has to see that blood does not accumulate in your legs at the expense of your head; but if you overdo it—if you stand long in a close crowd or a stuffy room—even this admirable system breaks down, and your brain, which is particularly sensitive about oxygen, runs short of blood. You "faint."

Here again science has only made a great discovery to be confronted with a mystery. We know that there are nerves from the muscles of the arteries to the spinal cord, and that the stopcock we have spoken of is regulated by a reflex nervous message from the cord; but how these unconscious elements of the human mechanism work so perfectly together we do not know. When we remember how densely ignorant of all these things men were only a few generations ago, we may be sure that much will be discovered by our children and grandchildren.

Every year, indeed, brings new discoveries of a remarkable kind. We have already noticed that certain chemicals called "hormones," of which we shall speak more fully immediately, are produced in various structures (ductless glands) of the body, and "posted" in the blood, as it were, for a distant organ. One of these "hormones" comes into play in connection with the blood. When a man is setting about some prolonged and strenuous exercise, nerve-messages go to stimulate certain glands near the kidneys called the adrenal or suprarenal glands. They supply one of these chemicals (*adrenalin*) to the blood, and it passes round the circulating system until it reaches the small arteries. It closes the cocks and shuts down the supply to organs which are not at the time required to be active, and thus ensures a fuller supply to those organs which are called into strenuous exercise.

When the blood has passed through the tissues—given up its nutriment and received the waste carbonic acid gas and the soluble nitrogenous waste—the blood turns back towards the heart. It passes into a new set of fine tubelets or capillaries, and these unite in the veins. The veins have thinner walls than the arteries, as there is now less pressure, but they have a remarkable series of valves along their course. The blood cannot flow back —cannot go wrong. You can actually trace one or two of the valves on the veins of your arms if you try to force the blood back to your fingers. Little knots stand up here and there. So the blood courses steadily back and is poured into the opposite side of the heart to that from which it started. It enters the right auricle, and passes on into the right ventricle; and the next beat of the heart sends it to the lungs, where it gives up its carbonic acid and gets a fresh supply of oxygen.

§ 7

How and Why We Breathe

The blood has many functions. It takes fluid food to the organs and, in its red corpuscles, it carries oxygen. It has also to bring away from the organs the waste products of their activity, the carbonic acid (carbon dioxide), which is got rid of in the lungs, and the soluble nitrogenous waste, which is got rid of by the kidneys. The work that is done in the various organs—such as the muscles, nerves, and glands—may be roughly compared to the work done in a simple steam-engine. Fuel must be supplied; then oxygen (the essential element of air) must be supplied to unite chemically with the fuel and convert the energy which is locked up in it into heat and active work. The stomach supplies the fuel. The lungs, like the blacksmith's bellows, supply the draught of oxygen. If we do not forget that in the animal body the chemical action is far more subtle and indirect than in the furnace, this may be taken as a useful simple view of what goes on.

Let us follow the draught of air into the lungs and the blood. We saw that, for the food to become available to the tissues, the blood-vessels have to become finer and finer until at last their walls are so thin that the nutritive material can pass through them. It is the same with the air-passages through which we breathe. The air enters by the nostrils; we will suppose, at least, that the reader is sensible enough to breathe always through the nostrils, and teach his children to do the same. In the nose there is a warming chamber, richly supplied with blood (and the supply automatically increases in cold weather), and there is a sort of sieve or filter (the hairs in the nostrils) for "screening" foreign bodies from the air. Dry air is also moistened in the nostrils. There is a mucous membrane in it which is most useful if you treat it reasonably; but if you treat it unreasonably—if you pack yourself with others into a moist, stuffy, unventilated railway-carriage or small room—it will get gored with blood and "boggy," and offer a good field for certain microbes, and—you will have a "cold in the head."

Behind the root of the tongue the air-way crosses the food-way and enters the windpipe or trachea. At this point there is the customary "door," automatically opening and shutting; and behind this delicate folding-door are the "vocal chords" which we use for speech.   The windpipe divides at its base into the two bronchial tubes, one for each lung, and there are ingenious arrangements for dealing with dust or microbes that have got past the sentinels in the nose.   There is a coat of mucus to inter-cept them (as the flypaper does flies), and there are countless microscopic lashes or cilia which bend and straighten rhythmic-ally, beating towards the entrance, and thus gradually push out the intruder.   If certain kinds of dangerous microbes settle on them, the glands pour out large quantities of mucus, and your lungs automatically blow it out at times—you have a "cold" and a "cough."

In the lungs themselves the bronchial tubes branch out into numbers of fine tubes as the arteries do, and each tube ends in a score or more of little air-chambers.   There are about six mil-lions of these minute chambers (each about 1-10 of an inch long) in the two lungs, and they are formed so as to give as much surface as possible.   If we could open them all out and piece them together, we should have a total surface a hundred times larger than our skin.   This is the wonderful contrivance evolved for bringing a large body of air into almost direct contact with the blood fifteen times a minute or more.   In a deep breath we can take in a whole gallon of air, and even a normal breath takes in two quarts.

Nerve Messages

But who attends to the working of this wonderful bellows while we are sleeping or are concerned with other matters?   It is, of course, another of those automatic mechanisms which have been formed in the animal body during millions of years of trial and sifting.   The lowest part of the brain, a part called the *medulla oblongata,* includes a nerve-centre which is sensitive to the carbonic acid in the blood and, stimulated by this, sends

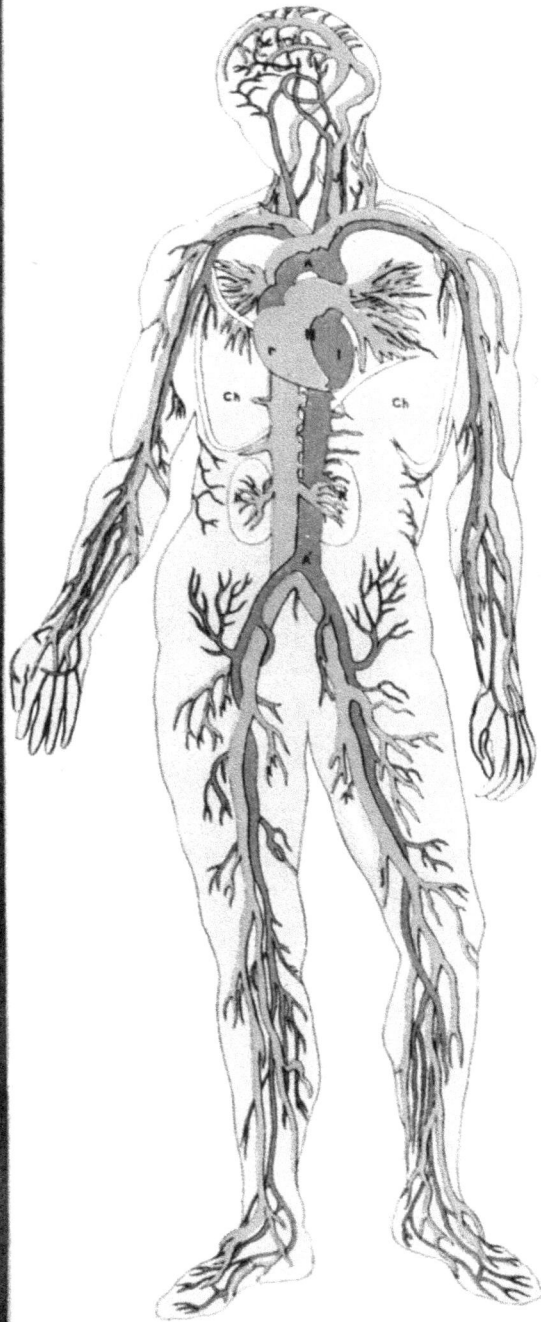

DISTRIBUTION OF THE MAIN BLOOD-VESSELS OF THE HUMAN BODY

The Arteries are shown in red, the Veins in blue. H, the heart; *l*, left side, *r*, right side. Arising from the left ventricle of the heart is the main artery, the dorsal aorta (A). The letter is put at some distance from the heart, near where the vessel gives off the branches (in red) for the head and arms (carotid and subclavian arteries), and at the point where it arches backwards and downwards to pass through the chest and abdomen, till at A^1 it gives off branches for the legs. The veins are represented running alongside of the arteries. Besides the dorsal aorta runs the great vein—the inferior vena cava—going to the right auricle. At K is represented the position of the kidneys and their renal vessels. L represents the pulmonary veins of the lung, the only veins with pure blood in them. J, jugular vein, coming down the neck. Ch, outlines of the chest.

automatic messages to the muscles of the ribs and the midriff, or diaphragm. At each intake of breath twelve pairs of muscles work harmoniously in expanding the chest, and then other muscles pull the "bag" together again and expel the air. But how can the air extract the carbonic acid from the blood in so short a space? All such difficulties are provided for in the body-machine. You breathe out only a fifth of the air in your lungs every time. The little air-chambers automatically close if, by a strong effort, you try to empty your lungs. The exchange of gases is going on all the time. If, on the other hand, the muscles are working hard, and need more oxygen, the increased carbonic acid in the blood stimulates the medulla, and nerve-messages from it rain upon the lung-muscles until you "pant for breath."

A man or woman engaged in sedentary work gets into a way of using the lungs to only about a tenth of their capacity. You understand the pale scholar and the anæmic girl in the cash-box. They provide too little oxygen, and the blood will not provide red corpuscles which are not needed. If such people will go for a good swinging walk, in air that is rich in oxygen, the blood will stream through their medulla, the nerve-centre at the base of the brain, and the lungs will open out.

The real "breathing" is, of course, deep inside the body. The little air-chambers have walls almost as thin as soap-bubbles, and a rich supply of similarly thin blood-vessels (capillaries) outside them. Through these thin walls the red corpuscles somehow extract the oxygen from the air, and the blood also gives off the carbonic acid. Then the blood streams back to the left chamber of the heart to be pumped through the body in the way we have described. When this blood finds itself in the thin-walled capillaries amongst the organs of the body, the red corpuscles yield their oxygen, and the blood returns to the heart with a new load of carbonic acid.

We have seen that this union of oxygen and food in the tissues may roughly be compared to what goes on in a steam-engine.   It enables the organs to work—to do the work we describe here—and it produces heat.   And in connection with this heat we employ, all our lives, wonderful mechanisms which even modern science has only partially mastered.

The blood must be kept at a temperature of (in a normal human body) about 98.4° F.   When the air sinks very low in temperature, we shiver, or stamp our feet, or rub our hands.   The shiver is an automatic warning to take exercise, to increase the combustion in the muscles.   When, on the other hand, the outside temperature rises too high, we get the stopcocks of our arteries, which are tightened on a cold day, now opened wide, to let the blood's heat escape by the skin.   If this does not suffice, automatic messages go from the nerve-centres to the millions of sweat-glands in the skin, and we "sweat."   To raise the temperature of the watery fluid so much heat has had to be extracted from the blood.   If the air is dry as well as warm, this mechanism is generally sufficient, but if we are in a "moist heat"—everybody knows how much worse it is than dry heat— the evaporation through the skin is checked, and the temperature of the blood rises until it may be too much for our brain.   Even cold moist weather is trying.   Our vitality is lowered in meeting it, and the cold-microbes get their chances to invade the body.

A wonderful mechanism surely!   But there seem to be unintended effects at times of these ingenious devices.   Take the "crimson flood" on a girl's cheek at some ugly word, or some word of praise, or some consciousness of guilt.   The stopcocks to the capillaries in her cheeks are opened wide, but we can hardly suppose that some nervous reaction was evolved for that purpose.   Sudden paleness is more intelligible.   The cheek blanches in the face of danger, because the stream of blood has been directed to the brain and muscles that may have to meet the situation, and such temporarily useless organs as the cheek have the supply cut off.

§ 8

The Large Machines of the Body

We may seem so far only to have concerned ourselves with organs which exist for the sake of other organs. That, in point of fact, is the nature of every organ in the "organism"; and indeed, it would be at least equally correct to say that bones and muscles, which one naturally thinks of as forming the greater part of the body, exist very largely for the purpose of digestion and respiration. Nutrition and reproduction are the oldest functions, the original functions, of the animal body. The elaborate skeleton, with its masses of muscles, has evolved to protect and minister to these fundamental activities.

Of the distribution of the two hundred bones and two hundred and sixty pairs of muscles which form the great bulk of the body little can be said here. A catalogue of the bones would be a list of unfamiliar terms; and a catalogue of the muscles would be almost an essay in Greek. It is in the development and minute structure of bone that modern physiology is chiefly interested. As is now generally known, the body begins its existence as a single cell—a microscopic speck of living matter surrounded by a membrane—and the development of the body is due to the repeated and rapid multiplication of this cell (the fertilised ovum or egg-cell), until countless millions are formed. It is a "cell-state": a commonwealth of millions of living, active units bound together into a harmoniously working organism.

As this "protoplasm"—the jelly-like matter which composes cells—is soft, the beginner may wonder how it can build up such structures as teeth and bones. To understand this, as far as we do understand it at present, we have to remember that, as the cells of the body multiply from the original egg-cell, they also separate into different classes. We get muscle-cells, nerve-cells, bone-cells, gland-cells, and so on; and they differ remarkably in structure from each other.

One contingent of these cells consists of the "bone-builders," and long before birth they begin to construct the supporting framework of the body. It is, of course, not bone at first. Frames of cartilage preceded bony frames in the course of racial evolution, and a cartilage-frame goes before bone in the development of the individual body. When the time comes, the bone-builders extract the lime-salts which have got into the blood with the digested food, and they use these in building up the bones. Sir Arthur Keith tells us that there are two million of these bone-builder cells at work in the thigh of a new-born baby, and that the number rises later to a hundred and fifty millions. They make the bones solid, then they change the interior into the light but strong texture with which everybody is familiar.

How is it that we feel no creaking, no jarring, or friction, of the two hundred and thirty joints by which our bones play upon one another? Here is another ingenious contrivance. A layer of cartilage remains over the end of each bone. It is dense, very elastic, and always well lubricated by one of the many remarkable automatic lubricating systems of the body. The cartilage cells themselves in this case are converted into lubricating fluid when they die!

The muscular system which moves the bones is the red flesh with which we are familiar in the butcher's shop. Everybody who has carved a joint, and knows the importance of cutting "against the grain," is aware that one of these large muscles of the ribs or limbs of a cow consists of muscular fibres packed closely in bundles. There are 600,000 fibres in a single muscle of man's arm, the biceps. Each fibre is composed of many fibrils, the seat of that power of contractility which we very little understand. The body-machine is still full of problems and mysteries for us. Three hundred years ago the courageous anatomists of

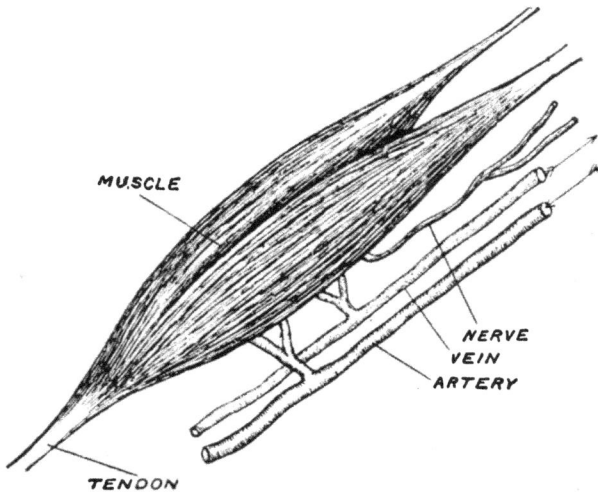

MUSCLE

NERVE
VEIN
ARTERY

TENDON

Reproduced by permission from Keith's "The Engines of the Human Body" (Williams & Norgate).

A DRAWING OF THE BICEPS OF THE UPPER PART OF THE RIGHT ARM, SHOWING ITS TENDON, ITS BLOOD-VESSELS, AND ITS NERVE

A tendon or sinew fastens a muscle to a bone; the artery brings oxygen and food-material for the muscle; the vein carries away carbon dioxide and waste; the nerve conveys the stimulus which provokes the muscle to contract. The biceps lies along the upper arm or humerus; its upper end is connected by two tendons with the shoulder-blade or scapula; its lower end is connected by a tendon with the radius, one of the bones of the lower arm; when the biceps contracts, becoming shorter and broader, as we can feel it doing, it raises the lower arm nearer the upper arm.

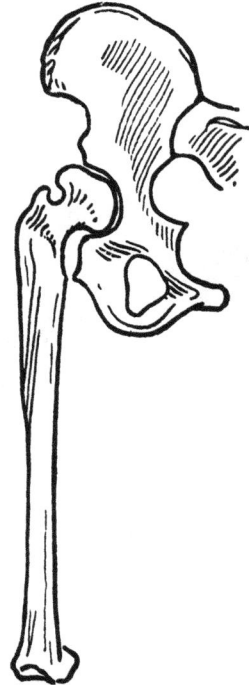

THE HIP-JOINT

The thigh-bone or femur is shown with its rounded head fitted into the socket (acetabulum) of the hip-girdle, which, in turn, is fixed to the sacral region of the backbone. This hip-joint is a good example of a "ball and socket" joint with a deep cup; the shoulder-joint is on the same principle, but with a shallow cup. The head of the thigh-bone plays in the cup furnished by the hip-girdle and has considerable freedom of movement, but much less than the arm, which plays in a shallow cup.

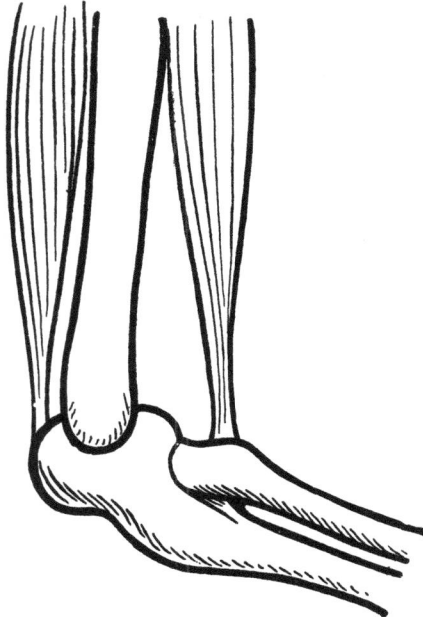

THE ELBOW-JOINT

The elbow is a fine example of a simple hinge-joint. The lower end of the humerus works on the upper end of the ulna, which bears an elbow process or olecranon, which prevents the arm being bent back. The biceps muscle, which is fixed above to the shoulder-blade, is inserted below on the radius, and bends the arm when it contracts. At the back of the elbow-joint is seen the triceps which straightens the arm when it contracts.

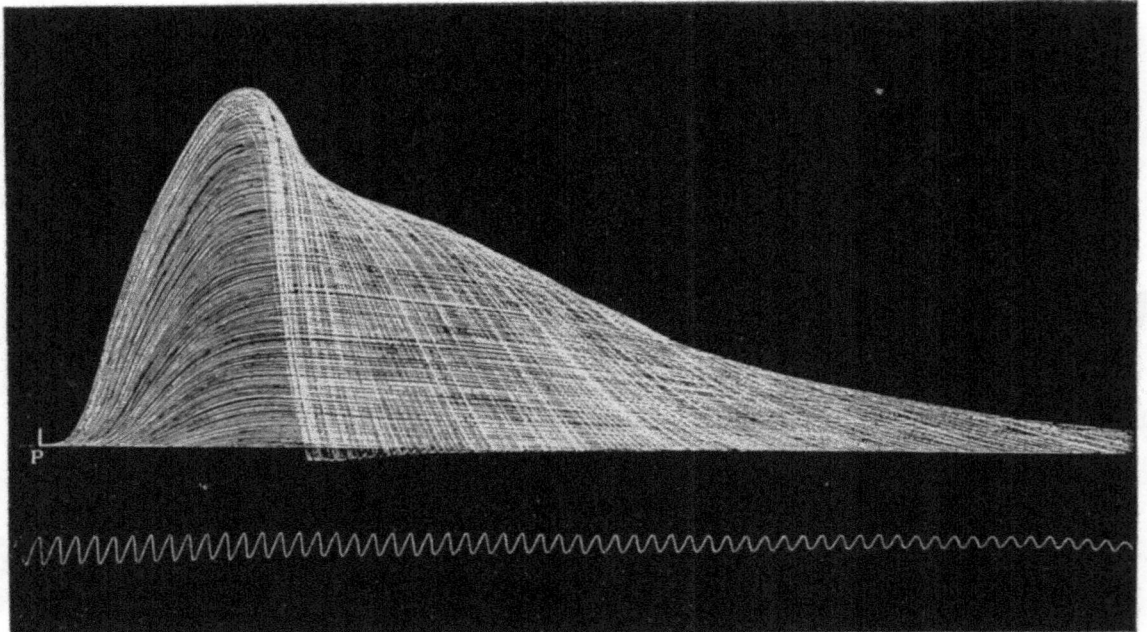

Reproduced by courtesy of Mr. John Murray from Halliburton's "Handbook of Physiology."

INFLUENCE OF FATIGUE ON THE CONTRACTION OF A MUSCLE

The muscle is made to write a series of curves, P being the point of stimulation; the lower separate wavy line is the time-tracing, the waves indicating hundredths of a second.   The point is this: For a time the contractions improve, becoming more and more vigorous, with higher and higher steep curves.   But by and by the contractions get less and less vigorous; they take longer and longer; the curves get flatter and flatter.   Finally the muscle ceases to contract at all.

the later Middle Ages began to make out the structure of the organs. Later came a generation which dissected the organs into tissues. Later still, as the microscope improved, the tissues were dissected into cells, and the whole life of the organism was resolved into the co-operative life of millions of these units. But we now know that the secrets of the life of the cell lie partly in the molecules which compose the cells, and these are beyond the range of the most powerful microscope. We must wait and be grateful for what we know. Science never rests. On the very day on which I am writing this page the press announces the discovery of a new microscope, which takes us at a bound deeper into the mysteries of living nature!

Meantime, science has shown us that the muscular system is an automatic living mechanism of the most wonderful kind. To every muscle the arteries bear their streams of food and oxygen, the muscle-cells select their diet, and the veins take away the waste-products. On every muscle there are also the fine endings of some nerve from (generally) the spinal cord, and at the proper moment a discharge along the nerve causes the whole mass of the cells or fibres in a muscle to contract simultaneously and lift the bone to which the muscle is attached. The nerve-impulse itself is slight. It is merely like a match set to the great energy stored up like powder in the muscle. But when we remember the number of muscles needed for a single harmonious action—we bring fifty-four into play at each step in walking, and there are about 300 muscles concerned when we walk—the delicacy of their adjustment, the precise degree of action needed in each, we cannot but marvel at the ceaseless regularity and correctness of this unconscious play of muscle and nerve and nerve-centre. We can say only that it is broadly and beautifully illluminated by the story of evolution—a slow advance during millions of years, during which every individual with a defect is sifted out and every improvement means longer survival in the struggle for life.

§ 9

THE NERVOUS SYSTEM AND THE BRAIN-CENTRE

The Nervous System

Most wonderful of all structures in the body-machine, and most difficult to understand, is the telegraphic system—the nerve-threads or "wires" and the central stations in the brain, the spinal cord, and certain other clusters of nerve-cells. A unified cluster of nerve-cells is called a ganglion or nerve-centre. In the simplest forms of life there is no nerve, no muscle, no mouth or stomach. The microscopic unit is one single cell—a bit of jelly-like living matter enclosed in a more or less definite membrane. Each and every part of it digests food, contributes to the movement, and is sensitive to the surroundings. In the course of evolution there arose larger organisms with *bodies,* with millions of cells bound together in harmony, *showing division of labour.* Some cells specialise on nutrition, some on reproduction, some on locomotion, and so on. Some of these cells specialise on sensitiveness, and thus arise nerve-cells. Then some specialise on one particular kind of sensitiveness, and there appear patches or pits in the skin, one sensitive to light, another to smells, and so on. Further advance unites these various centres by nerve-fibres, and at last a central telegraph station, a long tract of nerve-matter, connects up the various sense-centres and the muscles and glands. When a backbone is evolved, the main tract of central nerve-stations is enclosed in it; and, as life advances, the upper part of this "spinal cord" swells into a brain and is protected by a skull.

This interesting story of the evolution of the brain and sense-organs deserves to be told at greater length, but this slight outline may serve at present for our understanding of the essential nature of the nervous system. There is, as we said previously, a postal system and a telegraphic system in the body. Certain organs discharge certain chemicals (hormones) into the blood, and the blood delivers them to distant organs which are thus set to work. Obviously, this postal system would be too slow for the purposes of ordinary life, and so the telegraphic system is richly developed. Suppose that in bathing you tread upon a sharp stone. In a fraction of a second a nerve-thrill flashes from that part of your foot to a certain centre in the spinal cord, and a return thrill causes the muscles of the leg to contract, thereby jerking back the foot. In an animal far down in the scale like the octopus this nerve-message goes at about eighty inches a second; in the frog the speed has worked up to ninety feet a second; in man it reaches about four hundred feet a second.

In the case of man the nerve-message often goes on to ring a bell in the brain, as it were—to announce itself in consciousness —but the greater part of the body-machine is run, as we have seen, by automatic action, and we will first master this. We have spoken repeatedly of "reflex action." We mean by this nervous action without conscious effort. The message that goes to the brain or spine is automatically "reflected," along a different "wire," to the muscles or glands. When a piece of dust blows against your eyeball, one nerve sends some sort of thrill to a centre in the brain. Within a very small fraction of a second this message passes through a nerve-centre in the brain, and another thrill comes back to the muscles of the eyelids. Nearly the whole body is connected up, generally through the spinal cord, by an automatic nervous machinery of this kind. For the muscles of the head and face the nerve-centres are in the brain.

The nerve-cells (or neurons, as they are often called) have a cell body and outgrowing fibres or "wires." Each cell has two or more fibres running out to it, and these in most cases end in brushes of still finer threads. The nerve-cells are, therefore, particularly suitable for communicating with each other. In the brain and spinal cord especially each cell runs into a little brush or tree of fine fibres, and they interlock with each other. In the nerves that carry messages or commands to the muscles and glands many long fibres are bound up into bundles by a sheath. Inside each fibre there is a mysterious central channel, the axis cylinder, probably of a liquid nature.

What the real nature of a nerve-thrill is we do not yet know. It is accompanied by electricity, but it is not itself an electric wave, for such a wave would travel more than a million times faster than the nerve-message does. The nerves are also peculiar in the fact that they never get tired (as long as they are well supplied with oxygen), and physiologists have not been able to discover any definite chemical change in them. Even the production of carbon dioxide is questionable. Sleeping or waking, the wires are always alive, yet physiologists have not found that any heat is produced in connection with their activity.

The Brain-centre

It is otherwise with the masses of nerve-cells which make up the great brain-centres. Everbody knows that these grow tired, and must have a period of rest and recovery. Sleep is, however, still a puzzling phenomenon, and no theory of it can be regarded as satisfactory. All that physiologists are generally agreed upon is that the blood-supply to the brain is checked, and this lessening of the supply of oxygen (as to which the brain is particularly sensitive) lowers the vitality of the organs of consciousness. About the end of the first hour of sleep (which is the real "beauty-sleep") the brain-life is entirely suspended, and the blood is busy feeding the tired muscles. Some hours later more blood seems to return to the brain, and we get partial consciousness, uncontrolled by intellect, in the form of dreams. In a few individuals there may be, instead of a partial return of consciousness, an awakening of the power of automatic response to stimulations. They are apt to "walk in their sleep."

Our knowledge of the brain is now a special and formidable

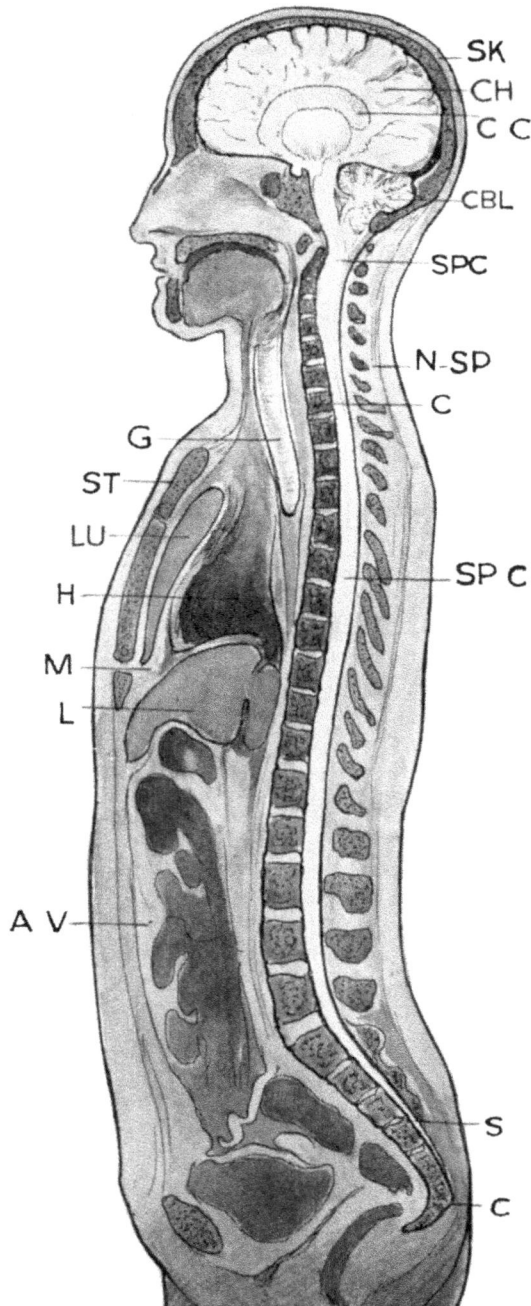

**VERTICAL SECTION THROUGH THE HEAD AND TRUNK
OF THE HUMAN BODY**

SK, the skull; CH, the cerebral hemispheres of the brain; C C, the corpus callosum, a bridge of nerve-fibres binding the cerebral hemispheres together; CBL, the cerebellum. SP C, the spinal cord; N SP, neural spines of the vertebræ; C M, a centrum or body of a vertebra; S, the sacral vertebræ fused; C, the coccyx, a fusion of post-sacral vertebræ; A V, intestine and other abdominal viscera; L, the liver; M, the muscular midriff or diaphragm, separating the abdominal cavity from the chest cavity; H, the heart; LU, the lung; ST, the sternum or breastbone; G, the gullet or œsophagus. From a specimen.

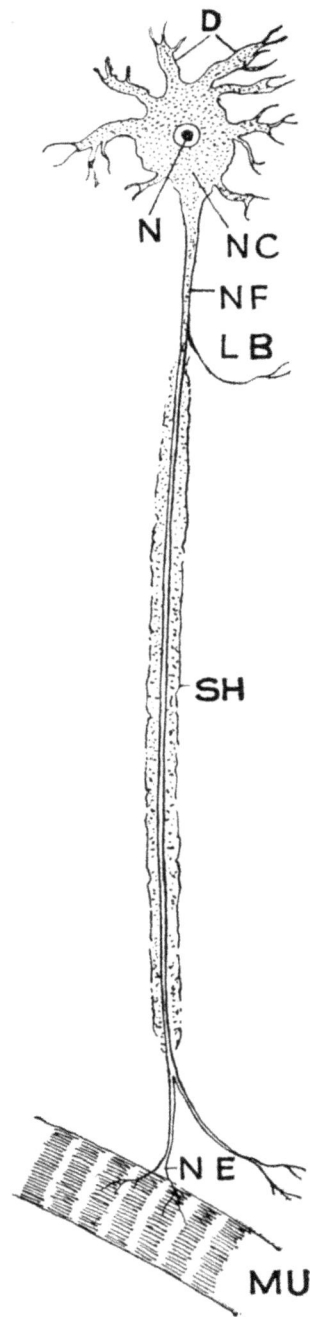

A SINGLE NERVE-CELL OR NEURONE
(After Stöhr.)

N is the nucleus of the cell; NC, the central-sell-substance or cytoplasm. The nerve-cell communicates with others by means of fine protoplasmic branches or dendrites (D). It gives off a nerve-fibre (NF) to a muscle (MU). This fibre has as its essential part an axis cylinder or core, surrounded by a medullary sheath (SH) of a fatty material; and outside this there is a clear membrane called the neurilemma. It will be observed that the medullary sheath is not developed at the origin or at the end of the nerve-fibre. A lateral branch of the fibre is shown (L B) and the ending (N E) on the muscle.

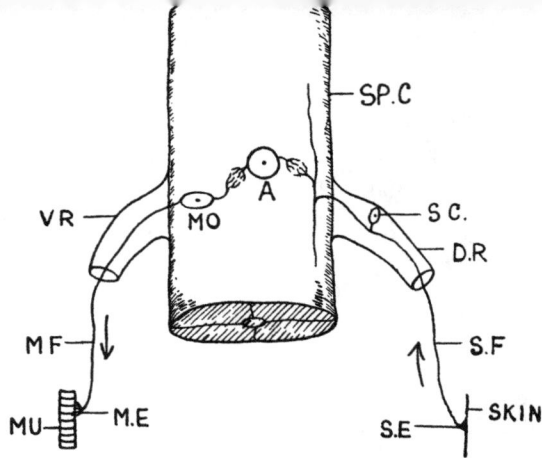

DIAGRAM ILLUSTRATING REFLEX ACTION IN MAN OR
ANY BACKBONED ANIMAL

From the sensory nerve-ending (S E) in the skin, a stimulus
passes up a sensory nerve-fibre (S F) to a sensory nerve-cell (S C)
in the spinal ganglion of a dorsal or afferent root (D R) of a spinal
nerve. The fibre, continued from the sensory nerve-cell, divides
in the spinal cord (SP C), and the message passes on to an asso-
ciative, intermediate, or internuncial nerve-cell (A). Thence it
is shunted to a motor nerve-cell (MO), from which a command
passes down a motor nerve-fibre (M F), issuing by a ventral or
efferent root (V R) of a spinal nerve. The motor nerve-fibre ends
in a nerve-plate (M E) on a muscle-fibre (MU), which is stimulated to contract.

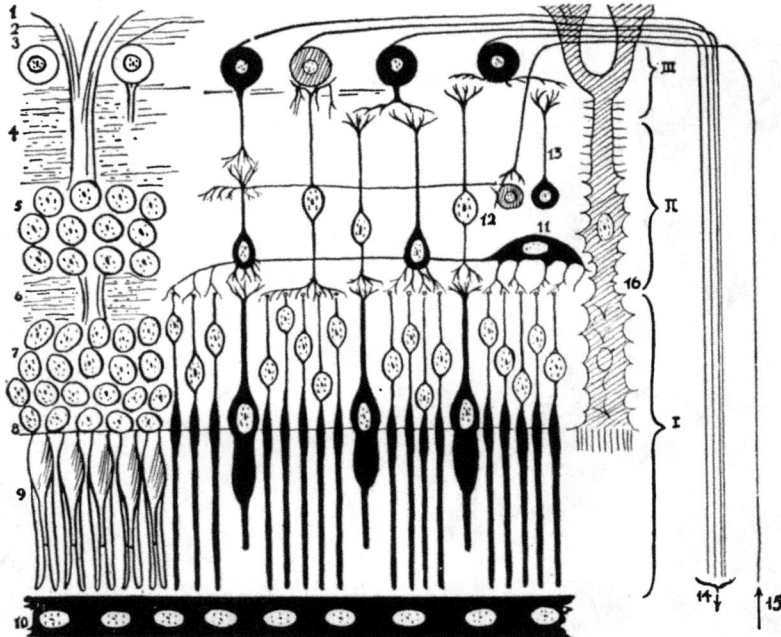

DIAGRAMMATIC CROSS-SECTION THROUGH THE RETINA OR PERCIPIENT
LAYER AT THE BACK OF THE EYE. (After Hesse.)

The figure gives some idea of the intricacy of this layer, which is not thicker than the
paper of this book.

1. Inner or anterior limiting membrane, next the vitreous humour in the cavity of the eye.
2. A branch of the optic nerve dividing up. 3. A layer of ganglion cells. 4. An inner layer of
nerve-fibres. 5. A layer of bipolar cells (so-called "inner granular layer"). 6. An outer
layer of nerve-fibres. 7. Layer of visual cells (so-called "outer granular layer"). 8. Outer
or posterior limiting membrane. 9. The rods (longer and thinner) and the cones (shorter and
broader). 10. Pigment layer of the retina. 11. Tangential cells. 12. Bipolar cells. 13.
An amacrine cell. 14. Centripetal fibres of the optic nerve. 15. Centrifugal fibres of the
optic nerve. 16. Muller's supporting cells. I, II, III, the three areas of nerve-cells in the
retina.

It is not in the least within the scope of this work to explain the minute structure of the
retina; the figure has been introduced to give some impression of the complexity of the vital
architecture. The essential fact is that the rods and cones somehow convert the pulses of the
luminiferous ether into stimulations of the fibres of the optic nerve.

branch of science; it will be referred to later when we come to deal with Mental Science.

"In some way that we do not understand, our personality is more bound up with our nervous system than with the rest of our body. Our quickness or slowness, alertness or dullness, cheerfulness or gloominess, reliability or fickleness, good-will or selfishness, are wrapped up—in our ordinary life inextricably—with our very wonderful nervous system. Some people believe that our inmost self uses the nervous system as a musician uses a piano, and compare the disorder of mind illustrated in the delirium of fever, or the decay of mental vigour in the aged, to disturbances or wear and tear in the instrument. Others think that the inner life of consciousness—feeling, thinking, and willing—is one aspect of our mysterious living, and that the physico-chemical bustle that goes on in the nervous system is the other aspect of the same reality. The two aspects are inseparable, like the concave and the convex surfaces of a dome; but no metaphor is of any use, the relation is quite unique.

"This is one of the oldest of riddles, and Tennyson made 'The Ancient Sage' say:

'Thou canst not prove that thou art body alone,
Nor canst thou prove that thou art spirit alone,
Nor canst thou prove that thou art both in one:

    .   .   .   .   .

For nothing worthy proving can be proven,
Nor yet disproven.'

"Yet three things seem to us to be quite certain: (1) Our nervous system is a scientific actuality that can be measured and weighed; it is complex beyond our power of conception, if only because of the millions of living units which it includes: it is the seat of an extraordinary activity which baffles the imagination. No theoretical view can stand that is subversive of the fundamental reality of our nervous system. (2) Even more real, however, if there are degrees of reality, is our inner life of consciousness, our stream of thoughts and feelings, desires and purposes. It is our supreme reality, for it includes all others, and no theoretical view can stand that is subversive of this reality. (3) But the third certainty is that organism and personality, body and mind, nervous metabolism and consciousness, are in the experience of everyday life interdependent. It is a relation, there is nothing to which we can compare it; if it is a unity, it is equally unique. We are mind-bodies or body-minds; sometimes we feel more of the one, sometimes more of the other." [1] That, however, as we have seen, will form the subject of a later chapter.

We may note here that it is a popular fallacy to suppose that all the contents of the skull are concerned with thought and feeling, or that a large head means a large capacity. The bulk of the matter in the cranium has nothing to do with thought. It is only a very thin rind or cortex of nervous matter, about a ninth of an inch thick on the average, covering the fore-part of the brain (from the top of the head to the base of the forehead) which is the organ of consciousness. But this precious cortex is an intricate structure made up of 9,200 million nerve-cells, and it is in man folded and creased so as to pack as much surface as possible within the limits of the human skull. Round this central area are the nerve-centres for controlling the muscles of the head, face, eyes, tongue, and the like; and the centres for receiving the reports of the eyes, nose, and ears are also in the brain. In a man who weighs 150 lbs., the nerve-cells of the brain-cortex would weigh about $\frac{1}{5000}$ part of the total, but this small part controls the whole.

At the back of the head is the cerebellum, or "small brain": the chief centre for co-ordinating the movements of the muscles so as to produce harmonious action. If it has been injured in a bird or a dog, the animal can no longer stand up or maintain a balance of movement. All day long the cerebellum must be re-

[1] Professor J. Arthur Thomson, *The Control of Life.*

ceiving countless messages from all parts of the body and directing our three hundred muscles to co-operate. It is entirely
automatic, yet no central telegraph station in the world is so busy
or so accurate. It also in some way maintains the tone of the
muscles.

Below the cerebellum is the medulla, which, as we saw, is the
organ for controlling the muscles of the chest that cause breathing. It has, however, much more work to do than this. It has
some control of the heart and blood-vessels, and it influences
movement in the alimentary canal from the salivary glands to the
small intestine. We must remember that these hind-parts of the
brain are the oldest. The cortex—the nervous matter connected
with mental life—is a later acquisition.

And the oldest part of all is the long cord of nerve-cells
which is enclosed in the spinal column. Along this are the various
centres for working automatically the great muscles of the trunk
and limbs and abdomen. Pairs of nerves leave it at intervals, and
all day long these are receiving messages and issuing orders. It
has an extraordinary power of automatic learning. Watch the
baby learning to adjust its muscular actions to its desires or feelings, or a girl learning tennis or typing. In a short time the
machinery will react promptly and perfectly to the stimulus. It
is through the spinal cord that the brain can influence movements
which are usually automatic.

We cannot discuss here how far the bodily features may
serve as indices of mental character, whether the face, the eyes,
the shape of the head or the hands—an interesting chapter on the
subject will be found in Sir Arthur Keith's little book, *The
Human Body*. There is no correspondence, he tells us, between
the functions of the various parts of the brain, so far as we yet
know them, and the overlying parts of the head to which "phre-
nologists" have assigned definite functions. Some day we may be
able to add to our knowledge of a man's character derived from

observation of the expression of his face, his words and actions.
"The day may come when by looking at the brain, or even at the
skull which encloses it, we shall be able to tell the capabilities of
a child or a man, but we have not yet reached that point." Neither
is it true that the lines of the palm of the hand can be "read" as
guides to the future: palmistry is childish make-believe

§ 10

THE ORGANS OF SENSE

The Organs of Sense

Another section of this work tells how our wonderful sense-
organs were slowly evolved, and it is enough here to observe that
they began as simple sensitive patches in the skin which, in the
course of millions of years, have grown into elaborate organs.
They are the sentinels of the commonwealth of cells. For ages
theirs was the vital function of locating food and announcing
danger. Now, in man, they are the chief channels of those
glimpses of nature which the mind unites in the marvellous
structure of modern science.

The skin, to begin with, is crowded with little organs of sense.
Nerves from the great centres branch out in every direction, and
the fine twigs at last end in sensitive bulbs underneath the skin.
The most numerous of these are for the purpose of announcing
"pain." We speak of pain as something in the body-machine
which we could very well spare, but a little reflection will soon
tell us that it is a most benevolent institution. It announces
some danger which, if it were not thus indicated by the ringing
of the bell in the brain, would not be noticed, and might become
fatal. Other little bulbs, especially numerous on the palm-side
of the fingers, minister to the sense of touch. Others feel cold,
and a different set experience heat. By careful testing, the reader
can find for himself that these sensations are localised in different

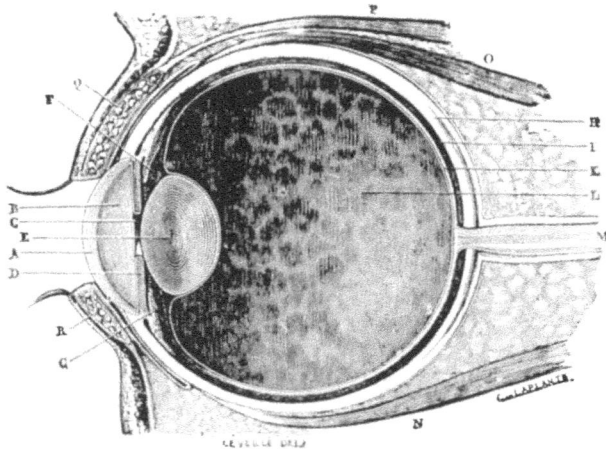

From "The Household Physician," by permission of Blackie & Son, Ltd.

REPRESENTATION OF A VERTICAL CUT THROUGH THE EYEBALL IN ITS SOCKET

In the figure, muscles (P, O, N) of the eyeball are shown. A is the cornea; it closes the front of the anterior chamber (B), which is filled with aqueous humour and the back wall of which is formed by the curtain of the iris (D). In the middle of the back wall is the opening of the pupil (C), through which is seen the lens (E). Behind the lens is the posterior chamber (L), filled with vitreous humour. Entering the eye from behind is the optic nerve (M), which is distributed to the retina (K). The posterior wall of the eye shows from within outwards the image-forming retina, the dark choroid with blood-vessels (I), and the firm protective sclerotic (H).

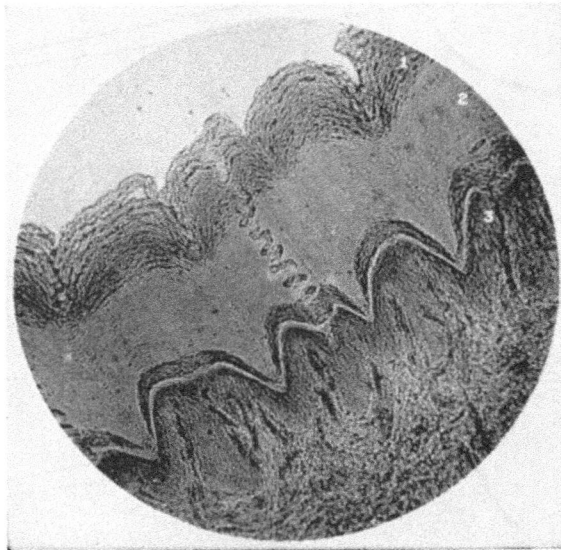

Reproduced by courtesy of Messrs. F. Davidson & Co.

SECTION OF HUMAN SKIN

1. To the outside with ridges is the horny layer of the epidermis (*stratum corneum*).

2. Then comes the second layer of the epidermis, the Malpighian layer (*stratum malpighi*); and traversing this is seen the coiled duct of a sweat gland.

3. Third comes the under-skin or dermis, the seat of many glands and blood-vessels. Its surface is raised in hillocks or papillæ. Into these there run blood-vessels and nerves.

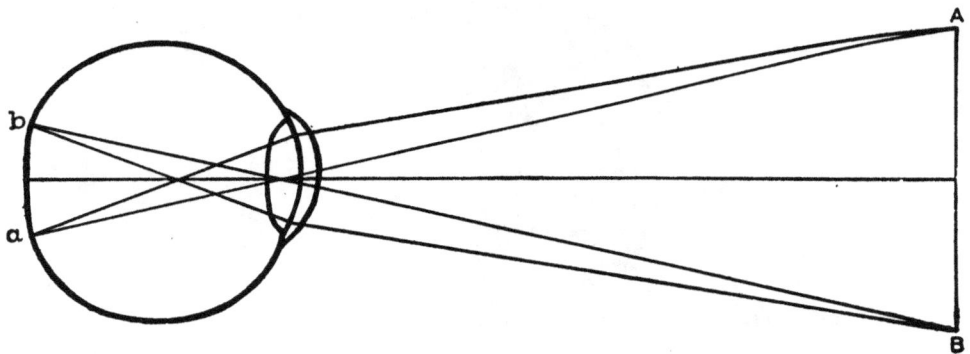

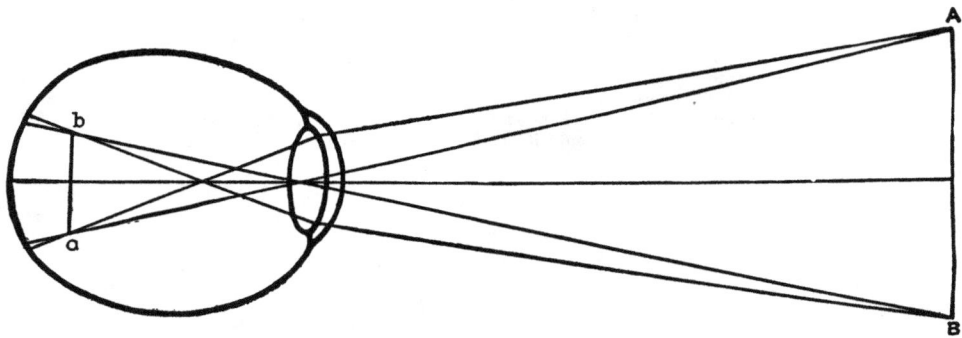

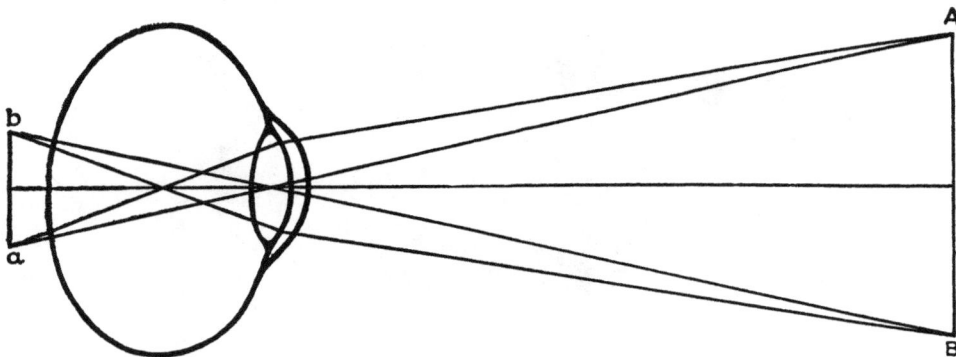

These three diagrams show the situation of the image in a normal eye, in the myopic or short-sighted eye, and in the hypermetropic or long-sighted eye. In the normal eye, the image a b is clearly produced on the retina at the back. In the short-sighted eye, the eyeball is too long and consequently the retina—or receptive tissue—is behind the clear image. In the long-sighted eye, the eyeball is too short and consequently the clear image is beyond or behind the retina.

areas on his arm. There are other nerve-endings again for the sense of pressure.

Other sensitive bulbs, which line part of the mouth, are the receiving organs for the sense of taste. Little oval bodies stand up like a close regiment of diminutive soldiers on the upper surface of the tongue. Each of the internal cells of these "taste-buds" ends in a hair-like process, and these processes touch the nerves which convey their particular stimulation to the brain. Probably different flavours are perceived by different nerves. The tip of the tongue is richer in the little bulbs that appreciate sweet things, while the back part is richer in the means of recognising bitterness.

Substances must be in a liquid form to announce themselves to taste. For the sense of smell, on the other hand, they have to be broken up into very fine particles like a gas. Nerves from the olfactory centres in the brain branch out in the membrane which lines the upper part of the nasal cavities, and this membrane includes numerous sensory nerve-cells which act as sentinels against dangers which announce themselves in the air. An odorous body is one which gives off minute particles of its matter into the air. The sense of smell was once of the gravest importance in the animal economy, and even in men it is so highly developed that they can detect a speck of musk diluted in eight million times as much air. A very strong offensive substance like mercaptan can be "sensed" even if there is only one grain to twenty-five trillion times as much air. In man, however, the sense of smell is degenerating, and many individuals have it very feebly.

The Sense of Vision

Most important of all the senses is that of vision, for nearly all the ideas of things in the mind of an ordinary person are visual images. The essential part of the mechanism of this sense is the eyeball and the nerve which goes from this to the sight-centre in the brain. The eye is a camera of a most remarkable description.

It is a roundish ball made of dense and strong fibrous tissue, opaque for five-sixths of its surface, but transparent in the one-sixth which bulges out in front, as the cornea. To the interior of the cornea, separated from it by a watery fluid, there is a delicate curtain which hangs over the transparent "window" in front and forms the variously coloured iris.

This curtain is a wonderful arrangement for adapting the eye to the intensity of light which falls on it. Fibres of muscles are so ingeniously distributed in it that it can almost close the opening in a strong light, or open it wide when the light is fainter. The "iris diaphragm" with which the photographer regulates the entrance of light into his camera is merely a poor imitation of it. Moreover, it contains pigment cells, which may be crowded when the light is strong or fewer in number when the eye wants as much light as possible. So we get the black eyes (eyes rich in pigment, to mitigate the light) of the southerner, the blue eyes (with little pigment) of the dweller in the darker northern lands, and every intermediate shade and combination of them.

Behind the circular window, the pupil, is the crystalline lens, which, unlike any lens that man can make, can be altered by fine muscles so as to focus itself for any distance. Other muscles and tendons are attached to the outside of the eyeball, and they automatically turn it in the direction we want. Some men of science have found many defects in the eye, and there are defects: but when one thinks of the unconscious agencies that have built up this wonderful camera, and work it automatically every moment of our waking lives, one is not disposed to cavil.

But the most wonderful part of it is the "sensitive plate" at the back of the eyeball. A semi-transparent membrane, which we call the retina, lines three-fourths of the interior of the eyeball (which is filled with fluid), and it is particularly developed at one spot, the real seat of distinct vision. On this "yellow spot" in each eye the rays of light form an inverted image of the object at which we are looking. The stereoscope enables us to understand how the images of the two eyes are blended, and how they enable us to see nature more perfectly.

Vision, as might be expected, is still very imperfectly understood. The retina is a very complex layer of delicate nerve-cells, in which certain parts that are known as "rods" and "cones" seem to be the essential elements. There seems to be chemical action, though whether there are three distinct chemicals for the three primary colours, or one chemical that breaks into separate colours, or what happens, we do not know. It is generally suspected that colour-vision is connected with one or more fine chemicals which may be lacking in "colour-blind" people. However that may be, the nerve-layer closes up at the back of the eye and, as the optic nerve, conveys the images of things in some way to the conscious centre. What precisely travels along the nerve we cannot say, but to imagine that an image or picture is conveyed is to imitate children who think that words travel along a telegraph wire.

The Sense of Hearing

The organ of hearing is not less remarkable than the eye. We have already seen that the external ear is, to use the cautious words of Professor Starling, probably of no use whatever. In cases where it has been cut off the sense of hearing was not affected at all. But it was useful and mobile in an earlier ancestor of man. From it, in any case, a narrow channel about an inch long, protected against adventurous insects by wax secreted by its glands, conducts the waves of sound to the real ear.

At the outer end of this passage the sound-waves beat upon a sensitive drum, the tympanum, a membrane of a most ingenious construction. This membrane must not have a period of vibration of its own. It must respond readily and immediately to every sort of wave that impinges on it. It is therefore so constructed that each part of it has a different period of vibration, and it is further "damped" by a little bone pressing against it on the other side. The pressure of air on the outside of the drum, which must alter with changes of pressure outside, is regulated by a channel (the Eustachian tube) running to it from the roof of the mouth.

Three little bones (the hammer, anvil, and stirrup) convey the vibrations of the drum to another drum, which is stretched across the entrance to the real ear inside the skull. As the waves of sound impinge on the tympanum and set it vibrating, the three little bones work together and repeat the vibrations on the second drum, the "oval window." Beyond this is a coiled shell which contains the real organ of hearing—a large number of hair-cells (the "organs of Corti") interlacing with the fine fibres of the auditory nerve. The vibration of the "oval window" agitates the fluid inside this organ, and the hair-cells communicate this movement to the nerves, which then convey it to the brain. Once more we have a mechanism full of ingenuity in every part, and brief descriptions of this kind are almost unjust to the various organs of our body; but to-day we should require a large volume to give an account of our knowledge of the brain and sense-organs alone. We have referred to the three small bones of the ear by which the waves of sound are conveyed to the inner ear. "The history of these bones is strange. The hammer was at an early stage of the evolution of mammals a part of the lower jaw; the anvil was the bone on the base of the skull, with which it articulated. When mastication and molar teeth were evolved in the ancestry of mammals, a new joint was formed in the lower jaw, and these two bones—the hammer and anvil—were taken into the service of the ear." [1]

§ 11

THE DISCOVERY OF HORMONES

Remarkable Discoveries about the Glands

A physiologist would class the different parts of the body as bones, muscles, nerves, and glands, and we have in conclusion

[1] Sir Arthur Keith, *The Human Body.*

STATUE SHOWING HUMAN MUSCLES

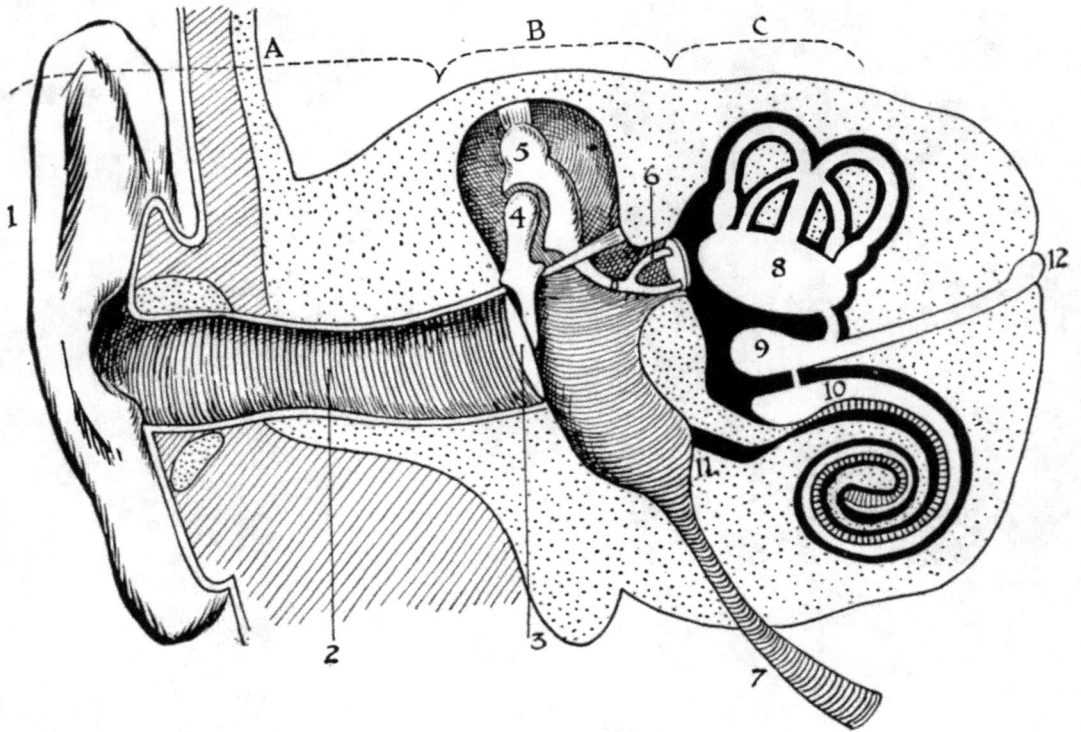

DIAGRAM OF THE HUMAN EAR. (After Hesse and Weber.)

A, the outer ear-passage; B, the middle ear; C, the bone (periotic) enclosing the inner ear.

1. The ear-trumpet or pinna, practically fixed in man and unimportant. In many mammals it helps to locate the sound. 2. The outer ear-passage with the drum or tympanum (3) running across its inner end. The drum vibrates when sound-waves strike it. 4, 5, 6. The ear-ossicles, hammer (malleus), anvil (incus), and stirrup (stapes), which by their movements transmit the vibrations from the drum to the inner ear. The "window" in the bony wall of the inner ear on which it abuts is called the fenestra ovalis. 7. The Eustachian tube, leading down to the back of the mouth; by it air can enter indirectly into the middle ear. 8. The larger chamber of the inner ear, called the utriculus, with three semicircular canals arising from it. They have to do with balancing and the like. 9. The smaller chamber or sacculus connected with the coiled cochlea, the essential organ of hearing (10), containing the organ of Corti. There is also shown the endolymphatic duct (12). 11. Another "window" in the periotic bone, the fenestra rotunda. The dark-coloured cavity is called the perilymph-space; it contains a fluid called perilymph; it is separated from the internal cavity of the ear by a membrane, within which there is endolymph. The dotted tissue is bone.

to say something about the last. We have already spoken of the myriads of tiny tubular glands which line the alimentary canal. Another mass of tubular glands make up the essential part of the kidneys. They really form a filter of a remarkable pattern. Arteries bring the blood to the kidney tubules, which are stimulated to action by the blood. Each—by vital action, not a mere physical process—takes out of the blood the fluid nitrogenous waste substances and a certain amount of water, and the tiny ducts connected with the tubules run together and carry the waste or urine to the bladder.

But the main interest to-day is in what scientific men call the "ductless" glands: glands which extract substances from the blood but do not pour their secretion into special channels. We have already mentioned a most interesting example in the suprarenal glands: two little bodies near the kidneys, each about the size of a segment of a small orange, which pour into the blood a "chemical messenger" (or hormone) for regulating the supply of blood to the various organs.

It is one of the most remarkable discoveries of recent years, that there are numbers of little glands entirely devoted to the manufacture of hormones. "If all the glands of internal secretion were rolled together they would form a parcel small enough to go into a waistcoat pocket, yet such a small mass can influence the working and growth of the whole body." In his interesting book, which we have already mentioned, Sir Arthur Keith, referring to the dispatch of secretion to the stomach, uses the following suggestive words: The secretin, or hormone, which acts as a missive " is posted in the nearest letter-boxes or capillaries in the duodenal wall and is carried away in the general blood circulation, which serves for all kinds of postal traffic. In a postal system where there are no sorters and which must be conducted by an automatic mechanism, letters or missives cannot be addressed in the usual way. Their destination is indicated, not by their inscription, but by their shape. The molecules of secretin may be regarded as ultra-microscopic Yale keys sent out to search for the locks of letter-boxes which they can fit and enter." They circulate round the body until they find their destination. "What is still more wonderful in this system is that the letter-boxes, or we may call them locks, have a positive attraction for the key missives which are destined for them." It was Professor Starling who named these messengers Hormones.

The thyroid glands—two little lobes on either side of the windpipe—are bodies of this nature which have attracted a good deal of popular attention of late years. The secretion formed is discharged straight into the blood stream, and for that reason they are called ductless glands or glands of internal secretion. They will be discussed at greater length in a later section of this work. Here it is enough to say that the chemical stuff, or "hormone," which they secrete increases the vitality of the tissues; it makes the tissues "greedy for oxygen," and the work goes on more briskly. Hence it is that decay or imperfect development of the thyroid glands leads to that state of bodily and mental feebleness which is called "cretinism," while the extract from the glands can be used for the purpose of "rejuvenation." This small organ, the thyroid gland, is necessary to the health and normal development of both body and mind, and this knowledge has been put to practical application in some cases with astounding results.

Near the thyroid glands are four small bodies—the "parathyroids." The function of these is not clear, but there is serious nervous trouble if they are removed. Then there is a "thymus gland," which seems in some way to prevent the sex-organs from developing too early. It is situated in front of the breast-bone, and must act by "postal service." The internal sex-organs themselves post a good many of these "hormones" in the blood. Everybody knows the striking difference between a normal and a castrated animal. The development of secondary sexual characteristics, such as antlers, seems to be largely stimulated by chemical messengers of this kind. One of the most interesting illustrations is in connection with the milk in the mother's breasts. How does the mammal mother come to have this rich development of her milk-glands just at the moment when it is needed? It has been discovered that, as soon as she becomes pregnant, the ovaries begin to discharge a hormone into her blood, which finds its way to the breasts and stimulates them. Probably the embryo itself also produces hormones which pass into the mother's blood, and serve a useful purpose up to the time of birth.

Finally, there is a remarkable and long neglected little body in the head, the "pituitary body," which is a rich laboratory of hormones. It controls the growth of tissues by stimulating them. When it is removed from an animal, the body becomes feeble and undersize. On the other hand, some rather unfortunate people have their pituitary body overgrown, or over-active, and they develop unpleasantly large faces, hands, and feet, or become "giants."

Such, as far as one can tell it in so brief a space, is the tale of the wonderful mechanisms in the body. Even the skin, which binds and protects this marvellous system of parts, is a remarkable organ when one has time to study it thoroughly. On the tender eyelids of a young child it is as thin as tissue paper, yet on the palms of some "horny-handed son of toil" it will produce protecting cells until it becomes an eighth of an inch thick. It is, moreover, rich in sweat-glands (which, as we saw, are most important for regulating the temperature of the body), lubricating or sebaceous glands, corpuscles for the sense of touch, and little pits in which take root the hairs which were once of great service to the body. Every internal surface also has its lining, or skin: tough where toughness is required, but so fine in the right places that gases and fluids can pass through it for breathing and nutritive purposes. "The proper study of mankind is man," said a great poet; and we may surely add that we know no more interesting study in the universe.

§ 12

Mind and Body

Before we leave the subject of this article a further word should be said.

The comparison of the body to an engine is very useful, but it is more than a little apt to lead us astray. For the body is living, and in higher animals, at least, there is a "mind" that counts. No one has succeeded in making clear the relation between mind and body, if there be a relation, but what we are sure of is that there are two aspects, two sides to the shield, the mental and the bodily. Just as a dome has its inner concave and its outer convex curve, inseparable from one another, two aspects of the same thing, so the living creature is a feeling, remembering, willing, and sometimes thinking being, just as really and truly as it is a feeding, moving, storing, and energy-transforming system. On the one side there is "mind," probably present even when it is not apparent to the observer; on the other side there is the routine of chemical processes which we call metabolism. Sometimes the living creature is more of a body-mind, sometimes more of a mind-body. We cannot solve the riddle: the mental or subjective and the bodily or objective activity are bound together in one. What we are quite sure of is that the ideal for the organism is a healthy body at the service of a healthy mind. Let us take an illustration of the influence of mind on body.

Emotions and Digestion

The famous physiologist of Petrograd, Professor Ivan Petrovitch Pavlov, was the first to demonstrate the influence of the emotions on the health of the system. Everyone knows that a good circulation and a good digestion make for cheerfulness, but the converse is also true. "A merry heart is the life of the flesh." The researches of Pavlov, Cannon, Carlson, and Crile have made it quite clear that pleasant emotions favour the secretion of the

Photo: Elliott & Fry, Ltd.

PROFESSOR E. H. STARLING

One of the leaders of modern physiology. He has made contributions of the highest importance towards an understanding of the chemical correlation or harmonious working of the body, notably by his study of internal secretions. Professors Bayliss and Starling have done much together.

Photo: Russell, London.

PROFESSOR W. M. BAYLISS

One of the leaders of modern physiology. He has made contributions of the highest importance to the study of internal secretions, digestion and the action of ferments, electric phenomena associated with vital processes, the production of heat in the body, and the regulative functions in general. His *Principles of General Physiology* is one of the great books of the science. Professors Bayliss and Starling have conducted many investigations together.

SIR JAMES YOUNG SIMPSON (1811–1870)

who introduced in 1847 the use of chloroform as an anæsthetic. He was Professor of Midwifery in the University of Edinburgh and was indefatigable both in research and in practice.

LORD LISTER

Joseph Lister (1827–1912), famous as the introducer of antiseptic methods into surgery. Following Pasteur, who showed that wounds went wrong with gangrene and the like because microbes or germs got in, Lister used a dressing, such as carbolic acid, which killed the microbes. Lister was Professor of Surgery in Glasgow, Edinburgh, and King's College, London.

digestive juices, the rhythmic movements of the food-canal which work the nutriment downwards, and even the absorption of what has been made soluble and diffusible. On the other hand unpleasant emotions, such as envy, and mental disturbances, such as worry, hinder digestion and the smooth working of the nutritive process.

The Cult of Joy

When the hungry man sees the well-laid table his mouth waters, but everyone knows that a memory or an anticipation will also serve to move at least the first link in the digestive chain. Professor Dearborn writes: "It is now well known that no sense-experience is too remote from the innervations of digestion to be taken into its associations, and serve as a stimulus of digestive movements and secretions." As was said of old time, "He that is of a merry heart has a continuous feast." When our joyous index is high, our digestion is good. As Dr. Saleeby has put it, freedom from care has nutritive value. It does not seem far-fetched to wonder if the joyousness of singing birds may not react on their remarkably well-developed capacities. We speak smilingly of our friend's "eupeptic" cheerfulness, but our smile is a little apt to become a materialism. We have to inquire whether our friend is not "eupeptic" because of his psychical success in the great task of happiness. The truth is that the mental and the bodily harmonies are the bass and treble of one tune.

The influence of mind on body finds a good illustration in the stimulation of the adrenal glands by strong emotion. Anger, which may be righteous, affects the production of adrenalin by the core of the adrenal glands, situated near the kidneys. The slight increase in this powerful "chemical messenger" or hormone, which the blood sweeps away, has numerous effects through the body. It constricts the smaller blood-vessels, and there is less blood in the peripheral and more in the deeper parts. It raises the blood-pressure, excites and freshens the muscles, adds to the sugar-content of the blood, increases the coagulability of the blood and so on. In short, the whole body is prepared for a fight, and all under the influence of what was to begin with a psychical event.

"Good news, psychical if anything is, may set in motion a series of vital processes, complex beyond the ken of the wisest. What is true of digestion is true also of the circulation. Wordsworth was a better physiologist than he knew when he spoke of his heart leaping up at the sight of the rainbow and filling with pleasure at the recollection of the daffodils dancing by the lakeside. There are facts which point to the conclusion that a gladsome mind increases the efficiency of the nervous system. Good tidings will invigorate the flagging energies of a band of explorers; an unexpected visit will change a wearied homesick child, as if by magic, into a dancing gladsome elf; a religious joy enables men and women to transcend the limits of our frail humanity."[1]

Healthy-mindedness

There is reason, then, to believe that emotion has its physical accompaniment in tensions and movements throughout the body, and in changes in the secretion of glands. There is a physiological reverberation of joy. But there must be more than this. The nervous system has a notable *integrative* or unifying function; it makes for the harmony of the bodily life. This function it may discharge the better if the psychical side is finding its due development. Thus it is well known that æsthetic emotion—delight in the beautiful—is very markedly a body-and-mind reaction, affecting the whole creature as a unity. It is practically certain that many people fail in health because they starve their higher senses and minds.

We venture to go further, under the conviction that physiology and psychology must join hands—as is suggested, indeed, by the name of the new science of psychobiology. The physiologi-

[1] Thomson, *The Control of Life,* 1921.

cal ideal is bodily health; its essential correlate is healthy-minded-ness. No doubt the invalid may have a vigorously healthy mind and the athlete a mind diseased, but, on the average, the two aspects of health must develop together. Hence the importance of mental dieting, mental gymnastics, mental rest, mental play, mental stores; though these must be sought as ends in themselves, not as aids to digestion!

BIBLIOGRAPHY

BAYLISS, W. M., *Principles of General Physiology* (1915). (A more advanced book for students, a standard work.)

FOSTER AND SHORE, *Physiology for Beginners.*

HARRIS, D. FRASER, *Nerves* (Home University Library).

HILL, ALEXANDER, *The Body at Work.*

HUXLEY, T. H., *Elementary Lessons in Physiology.*

KEITH, SIR ARTHUR, *The Engines of the Human Body* (1919).

KEITH, SIR ARTHUR, *The Human Body* (Home University Library). (A very interesting little book on the history of the human body.)

McDOUGALL, W., *Body and Mind* (1911).

McKENDRICK, J. G., *Physiology* (Home University Library).

POPE, A. E., *Essentials of Anatomy and Physiology* (1922).

THOMSON, J. ARTHUR, *The Control of Life* (1921); and *Secrets of Animal Life* (1919).

XI
HOW DARWINISM STANDS TO-DAY

HOW DARWINISM STANDS TO-DAY

Universal Acceptance of the Evolution Idea

WHEN people speak of Darwinism they sometimes mean the general idea of evolution—that the present is the child of the past and the parent of the future. Now the evolutionary way of looking at things has certainly been confirmed by the progress of science and is almost unanimously accepted by competent judges to-day. This horse that gallops past on the tiptoe of one digit on each foot is the natural outcome of an ancestral stock of small-hoofed mammals that used to plod about in the Eocene meadows, with four toes on each fore-foot and three and a vestige on each hind-foot. This bird that flies past is the descendant of such an old-fashioned type as the Jurassic Archæopteryx—an archaic bird with teeth in both jaws, a long tail like a lizard's, and a sort of half-made wing. And this first-known bird must be traced back to an ancestry among the extinct Dinosaur reptiles, though the precise pedigree remains hidden in the rocks. These reptiles must be traced back to certain primitive amphibians, and these to certain old-fashioned fishes, and so on, back and back, till we lose our clue in the thick mist of life's beginnings. If this is Darwinism it stands more firmly than ever, except that we are more keenly aware than in Darwin's day of our ignorance as to the origin and affiliation of the great classes. But, frankly, the only scientific way of looking at the present-day fauna and flora is to regard them as the outcome of a natural evolution. In a previous chapter this statement has been justified.

The Factors in Evolution

But "Darwinism" is more properly used, in a stricter sense, to mean Darwin's theory of the factors in evolution. If birds sprang from Dinosaur reptiles, if the modern horse is the descendant of Eohippus, which was about the size of a fox-terrier, how did the gradual transformation come about? There were many evolutionists before Darwin, and some of them propounded theories as to the factors in the age-long process. But Charles Darwin and his magnanimous fellow-worker, Alfred Russel Wallace, thought out a coherent theory of the factors—a theory that fitted the facts so reasonably well that it soon won the conviction of a large body of naturalists. The essence of the Darwinian theory is in the two words Variation and Selection; and Darwin stated it in a couple of sentences: "As many more individuals of each species are born than can possibly survive, and as, consequently, there is a frequently recurring struggle for existence, it follows that any being, if it vary however slightly in any manner profitable to itself, under the complex and sometimes varying conditions of life, will have a better chance of surviving, and thus be naturally selected. From the strong principle of inheritance any selected variety will tend to propagate its new and modified form."

The Essence of Darwinism

This is, however, too terse a statement. It requires some disentangling and expansion. *Proposition I.*—Variability is a fact of life. Offspring are usually somewhat different from their parents and from the other members of the family. Some of these variations make for success—success in getting food, in avoiding enemies, in securing mates, in giving the next generation a good start, and in other ways. Individuals that have varied in a profitable way will succeed better than those that have varied in the opposite direction, and better than those that have not varied at all.

Proposition II.—If the individuals that have varied profitably get the reward of their superiority, and if the individuals that have varied unprofitably, or not at all, are handicapped by their inferiority, this will have an effect on the character of the stock, or race, or species, *provided* that the novel peculiarities are hereditarily entailed on successive generations. If the individuals with profitable peculiarities (let us say, *plus* variants) are consistently favoured, and if their virtues are consistently handed on, their type will come to be that of the race. Whereas those with unprofitable peculiarities or none at all (let us say, *minus* variants) will be weeded out and will gradually disappear. Professor R. C. Punnett has calculated that, "If a population contains .001 per cent. of a new variety, and if that variety has even a 5 per cent. selection advantage over the original form, the latter will almost completely disappear in less than a hundred generations."

Proposition III.—But there cannot be sifting or selection without a sieve, and that is to be found in the struggle for existence. Living creatures are hemmed in by limitations and confronted by ever-changing difficulties. There is a tendency to over-population; circumstances are changeful; the vigorous creature is a hustler. There is struggle for food, for foothold, for self-preservation, for mates and for family well-being—indeed, for luxuries as well as for necessaries. There is struggle between fellows of the same kind, for a hungry locust may devour its neighbour and even the Amœba may be a cannibal. There is a struggle between foes of quite different kinds, between the grazing herd and the marauding carnivores, between the kestrel hawk and the nimble field-voles. There is struggle also between living creatures and their inanimate surroundings, the struggle against cold and heat, against wind and wave, against drought and flood. Subtle beyond description and almost ceaseless in its operation is nature's sifting, which Darwin called Natural Selection. In domestication and cultivation it is Man who fosters and eliminates; in nature the same kind of transformation as the breeder and the gardener effect is brought about by the struggle for existence.

Darwinism in Process of Evolution

These three propositions express the gist of Darwinism, and the question before us is, *How Darwinism stands to-day*. Before trying to answer this difficult question, we may point out that it would be a sorry business if Darwinism stood to-day as it was left by Darwin. He knew well that he had only begun to solve the problem of organic evolution; he looked forward with clear eyes to changes that the progress of science would enforce. It would be a terrible contradiction in terms if an evolution theory did not itself evolve! The marvel is, not that it is necessary to make some changes in what Alfred Russel Wallace so generously called "Darwinism," but rather that so much of Darwin's doctrine stands firm, four-square to the winds.

Another preliminary note is unfortunately necessary, that it is quite illegitimate to infer from our dubiety in regard to the *factors* of evolution any hesitation as to the *fact*. Our frankness in admitting difficulties and relative ignorance in regard to the variations and selections that led from certain Dinosaurs to Birds cannot be used by any fair-minded inquirer as an argument against the idea of evolution. For how else could Birds have arisen? As Wallace said in 1888, "Descent with modification is now universally accepted as the order of nature in the organic world." But the question before us is this: What, as regards the factors in evolution, have been the changes since Darwin's day?

§ 1

The Three Problems of Evolution

There are three great problems before the evolutionist: (1) What is the origin of the new? (2) What are the laws of inheritance? (3) What are the sifting methods that operate on

Photo: Becker and Maas.

PROFESSOR WILLIAM BATESON, F.R.S.

One of the most distinguished of the experimental evolutionists, he has made fundamental contributions to our knowledge of Mendelian heredity and of variation. He has confirmed Mendel's theory and added important elaborations. He has shown that discontinuous variation or mutation is of frequent occurrence. He was President of the British Association on its visit to Australia in 1914.

Photo. Rischgitz Collection.

GREGOR MENDEL, ONE OF THE FOUNDERS OF THE
SCIENTIFIC STUDY OF HEREDITY

Gregor Johann Mendel (1822–84), the son of well-to-do peasants in Silesia, became a priest in 1847, studied physics and natural science at Vienna, from 1851 to 1853, eventually became Abbot of Brünn. In the garden of the monastery he made experiments with peas, hawk-weeds, and bees; and published in 1865 what must be regarded as one of the greatest of biological discoveries. It was practically lost sight of till 1900. Regarding " Mendel's Law," Professor Bateson says· " The experiments which led to this advance in knowledge are worthy to rank with those that laid the foundation of the atomic laws of chemistry."

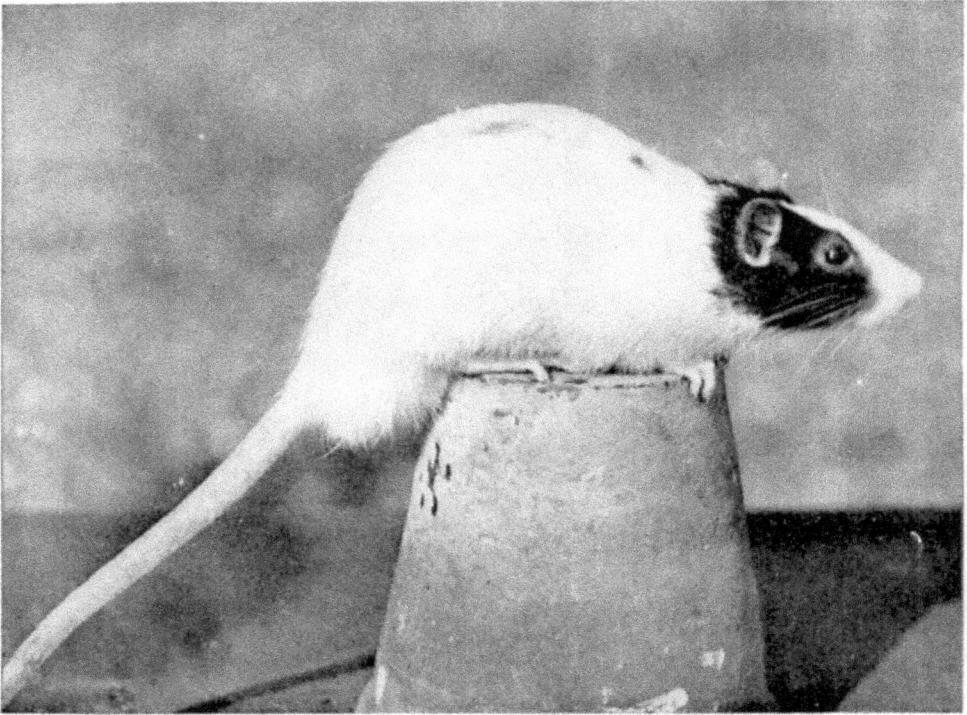

A RAT-BREEDER'S TRIUMPH

A "Dutch-marked" cross between a black and a white rat bred by Mr. H. C. Brooke. It shows a probably unique symmetry of markings. The black and white rats of the fancier are both derived from the common **brown rat** (*Rattus norvegicus*) and have nothing to do with the wild black rat (*Rattus rattus*).

VARIATION IN THE MAGPIE MOTH

This common moth, *Abraxas grossulariata*, which ranges from Britain to Japan, and is famous for its extraordinary number of variations in colour and markings. Such slight differences as those between the type A and the variety B (*lacticolor*) illustrate the minute variations which form part of the raw material of evolution.

the raw materials provided, and determine survival? In other words: what are the originative factors, what are the laws of entail, and what are the directive or sifting factors?

Evolution depends on new departures, peculiarities, idiosyncrasies, divergences, freaks, sports, a little more of this, a little less of that—in short, organic or constitutional changes. These are technically called variations and mutations. In other words, evolution—whether progressive or retrogressive—depends on the emergence of novelties. When there are no novelties there can be no evolution. The Lamp-shell, Lingula, seems to have remained stagnant for many millions of years—a fine creature, but icily perfect.

Heredity is the relation of organic continuity between successive generations, the living on of the past in the present, the flesh and blood linkage between an individual and his forbears on the one hand, his offspring on the other. The individual is like a lens into which rays from parentage and ancestry converge, from which they diverge again to the progeny.

Heredity is the reproductive relation which secures that like *tends* to beget like, and yet seldom does. Some peculiarities of an individual are heritable, others are not; longevity is readily entailed, but genius is not; deaf-mutism is very transmissible, but a very brown Anglo-Indian father has a peach-blossom complexioned daughter. Thus, if we think early, we see that heredity is not so much a factor in evolution as a *condition*. There would still be heredity though evolution stopped; but there can be no evolution without heredity. For heredity implies that the gains of the past can be capitalised; and, contrariwise, that individual losses need not involve racial bankruptcy. A man who has lost an eye may be assured that his son will have two, even if the mother is single-eyed as well.

What are called variations and mutations in biological language are the organism's experiment in self-expression, and these are the raw materials of progress. Granted raw materials, and granted their continuance, something more is needed—their sifting. As we have said in a previous article, the process of evolution is a long drawn-out process of testing all things and holding fast that which is good. The variations or novelties are the qualities to be tested; the struggle for existence, which includes the organism's endeavours, is the sieve that tests; *heredity secures the holding fast of what has proved good.* To employ a metaphor which has the defect of triviality, the variations are the ever-fresh hands of heredity cards that are given to the organism to play with; the organism uses these in the struggle for existence —with its strange mixture of active endeavour and fortuity. But when the organism with a good hand—a persistently good hand—becomes eventually tired and vacates its chair for a successor, it hands on its luck, and its cunning, too. Thus the essence of Darwinism is that nothing succeeds like success.

§ 2

As regards Variations

The fountain of change is even more copious than Darwin supposed. What is so clear in regard to pigeons and poultry, dogs and horses, that they are continually producing something new in their humanly controlled breeding, finds abundant illustration in wild nature. There are conservative types, it is true, which persist in a well-poised organic equilibrium, but in many cases there is flux. Outlying variants link one species to another. When the novelties or variations are registered statistically they often form what is called the Curve of Frequency of Error, which means that the number of variants of any particular magnitude will be in inverse proportion to the amount of the deviation from the mean. If the mean stature of the population be 5 ft. 8 in., there will be (as Alfred Russel Wallace points out) in 2,600 men, taken at random, one of 4 ft. 8 in. and one of 6 ft. 8 in., twelve of 5 ft, and about twelve of 6 ft. 4 in. In fact, there will be equal numbers at equal distances on each side of the mean, but the great majority of the deviations will be not far from the mean.

Definiteness in Variation

Since Darwin's time evidence has accumulated which shows that variations are more definite than used to be supposed. The palæontologists, who work out long series of fossils, bring forward cases of what looks like steady progress in a definite direction. There is a striking absence of what one might call arrows shot at a venture. It looks as if the occurrence of the new were limited by what has gone before, just as the architecture of a building that has been erected determines in some measure the style of any addition. An organic new departure will tend to be more or less congruent with what has been previously established. In post-Darwinian days the element of the fortuitous has shrunk.

Discontinuous Variations

Darwin was much interested in sports or freaks, such as the sudden appearance of a dwarf or a giant, a hornless calf or a tailless kitten, a white blackbird or a weeping ash, a thornless rose or a stoneless plum, a "wonder-horse" with its mane reaching the ground, or a Japanese cock with a tail six feet long. But Darwin did not venture to attach great importance to these brusque novelties, or discontinuous variations, first because he thought they were of rare occurrence, and second because he thought they would be speedily averaged off in the offspring of a sport which had paired with an ordinary individual. He did not know what his contemporary Mendel proved, that when a pure-bred tall pea and a pure-bred dwarf pea are crossed the offspring are all *tall*.

Now one of the great changes that has come about since Darwin's day is a recognition of the frequency of discontinuous variations, by which we mean sudden novelties which are not connected with the type of the species by intermediate gradations. We may think of the white crow or the weeping willow. The Proteus leaps as well as creeps. Especially through the investigations of Professor William Bateson and Professor Hugo de Vries, it has become plain that changes of considerable magnitude may occur at a bound. When the new character that suddenly appears, such as a Shirley Poppy or a short-legged Ancon Sheep, has a considerable degree of perfection from its first appearance, is independently heritable to the offspring, and does not blend or average off, it is called a *Mutation*. Professor de Vries writes: "The current belief assumes that species are slowly changed into new types. In contradiction to this conception the theory of mutation assumes that new species and varieties are produced from existing forms by sudden leaps. The parent type itself remains unchanged throughout the process, and may repeatedly give birth to new forms. These may arise simultaneously and in groups, or separately at more or less widely distant periods." This was strikingly illustrated by the sporting Evening Primrose (*Œnothera lamarckiana*), a species of North American origin, which de Vries found at Hilversum in Holland, and which proved to be in a very changeful mood. Almost all its organs were varying, as if swayed by a restless internal tide. It gave rise abruptly to numerous new forms which bred true. It illustrated species in the making.

Darwin found the raw material of evolution in small fluctuating variations, which are no doubt of frequent occurrence. Since Darwin's day it has become not only possible but necessary to attach much importance to discontinuous mutations. The contrast was aptly illustrated by Sir Francis Galton, who compared the varying organism to a polyhedron (a solid body with many faces) which can roll from one face to another. When it settles down on any particular face it is in stable equilibrium. Small disturbances may make the polyhedron oscillate, but it always returns to the same face. These oscillations are like

Reproduced by courtesy of Messrs. Methuen & Co. from "A History of Birds," by W. P. Pycraft. (After a drawing by G. E. Lodge.)

A FACTOR IN THE STRUGGLE FOR EXISTENCE (PEREGRINE FALCON ATTACKING A ROOK)

The Peregrine Falcon, which has been described as "the most powerful bird for its bulk that flies," preys largely on other birds, which it attacks during flight. The Falcon's aim is always to get higher than its quarry; it then "stoops" from above, killing not by force of impact but by the grip of its strong talons. As in many birds of prey, the female is larger and stronger than her mate and can hunt larger game.

Darwin's fluctuating variations, but the comparison breaks down inasmuch as the living creature may be, as it were, fixed in one of its oscillations, so that the variant makes a fresh start. Greater disturbances of the polyhedron may make it roll over on to a new face altogether, where it comes to rest again, only showing once more the minor fluctuations around its new centre. The new position corresponds to what is now called a mutation. Studies in inheritance have shown that these mutations have great staying power; they reappear persistently and intact in a certain proportion of the descendants. They are not liable to be swamped by intercrossing, as Darwin supposed. The curious fact is that the hereditary entailment of the fluctuating variations, which Darwin almost took for granted, requires more demonstration to-day than does the hereditary entailment of mutations.

§ 3

Variations and Modifications

Under the influence of persistent exercise, such as dancing, the muscles of the legs increase in size, and the tendency to increase may spread in an interesting way to other parts of the body. Long-continued exercise of white rats increases the weight of the heart, kidneys, and liver, on an average about 20 per cent. Water-snails reared in cramped surroundings grow up dwarfish. Goldfishes kept in the dark for three years become totally blind. If the wan pigmentless Proteus from the Dalmatian caves be exposed to light it becomes black, and the eggs laid by individuals kept in the light develop into dark larvæ. Prolonged pressure on a particular part of the skin often produces a thickening or callosity. The colours of birds' feathers are sometimes affected by the food they eat, as is well known in the case of canaries and parrots. The stomach of the herring-gull changes its character according to the diet—whether it be fish or grain.

Now all these changes are technically called "modifications";

they are directly induced *in the individual lifetime* by peculiarities, in habits and surroundings, including food. They are also called "acquired characters"—a very unfortunate term. *They are impressed from without,* whereas true variations and mutations *are expressed from within.*

Modifications are indents or imprints, variations are outcomes. According to the evolution theory of Lamarck, which Darwin accepted in some measure, the characters of a race may slowly change through the cumulative inheritance of the modifications which individuals acquire as the result of peculiarities in use and disuse, and in surroundings. A cave animal is blind, according to Lamarck, as the result of ages of living in darkness, during which the eyes have suffered from disuse. The modern Darwinian would point to the fact that constitutional or germinal variations in eyes are common. Variants with weak eyes and with a bias in that direction would naturally seek out caves. The giraffe has got a very long straight neck because of the cumulative result of generation after generation of stretching up to the branches of the acacia-trees. With certain provisos Darwin inclined to accept this view as supplementary to his own. But the modern Darwinian would point to the fact that constitutional or germinal variations in the proportions of different parts of the body are common. Giraffe variants in the direction of a long neck would prosper, and would become the leaders of the race. Long noses often run in families, but the length of the nose is not due to the vigour with which generations have used the handkerchief.

No one doubts the reality of modifications: one has only to look at the tanned skin of the African explorer. But what is doubtful is that a modification can be passed on from the individual that acquires it to his offspring—passed on as such or in any representative degree. The modification may be very important, even life-saving, for the individual, but unless it can be transmitted it is not in any direct way important for the race.

The scepticism as to transmission of bodily modifications was focussed by Sir Francis Galton and by Professor August Weismann; and many would say that one of the great changes in Darwinism since Darwin's day has been the abandonment of belief in the Lamarckian postulate of the transmission of modifications. There are some difficult cases, however, which suggest that biologists must not be in a hurry to shut out the possibility of such transmission. Admitting a few difficult cases, we can only record our impression that the available evidence indicating a transmission of "acquired characters" as such or in any representative degree is very inconclusive. But this would not be admitted by such a distinguished zoologist as Professor E. W. MacBride; and the scientific outlook should be that of an open mind, associated with an eager search for more facts.

Those who are unfamiliar with the subject often ask how a race could make progress at all if acquired characters were not transmitted from generation to generation. The answer is that the changes which make for racial progress are variations and mutations—*arising from within,* from disturbances and rearrangements, permutations and combinations, in the germ-cells from which new individuals arise. In 1796 the utmost speed of the trotting horse was stated at a mile in 2 min. 37 sec.; in 1896 at 2 min. 10 sec. Does it not follow that the trotting horse has been improved by the transmission of the results of the systematic training in trotting? It is certain that this conclusion does not follow from the available evidence, which points to the conclusion that the improvement in speed has been mainly due to the selective breeding of constitutionally swift horses. The trotter is born, not made.

It should also be understood that modifications may reappear, *not because* they have been transmitted, but because the conditions which originally brought about the change may still persist and produce the same effect on the offspring. And as to the inheritance of disease, this is apparently confined to constitu-

tional diseases which are due to disturbances in the germ-
cells. Diseases due to peculiarities of occupation or diet are
not transmitted as such, though an unborn offspring may
be poisoned before birth, or even infected with some disease
microbe.

Another common misunderstanding must be cleared up,
namely, the idea that if peculiarities directly induced by improvements in human "nurture" (surroundings, food, and habits) are
not handed on to the offspring, then such improvements are not
of great importance. But if the beneficial results of improved
function and environment are not as such transmitted, it becomes
all the more urgent that they should be reimpressed on each
successive generation. If they are not entailed, then it is all
the more important that they should be *re-acquired*. Moreover,
these ameliorations of "nurture" (in the wide sense) may serve
as the liberating stimuli that encourage the unfolding of new
variations of a useful sort. Besides, it has to be borne in mind
that, although the direct effects of fresh air, exercise, good food,
beautiful surroundings, pleasant work, and the like, may not
be transmitted as such or in any representative degree, they may
increase the general vigour of the next generation, and will certainly do so when the mother influences the offspring before
birth—an influence which is not in the strict sense part of the
inheritance. Given a constitutional taint or weakness, it may
be counteracted by suitable "nurture," but that will not make
it disappear from the inheritance. It will crop up in a later
generation if it gets a chance. In breeding animals and cultivating plants there seems to be no use working with individuals
showing advantageous *modifications;* the only hope is to select
from among advantageous *variations* or *mutations*. Finally, it
should be noted that if advantageous modifications are not entailed, which may be a matter for regret, the same non-trans-
mission will hold in regard to disadvantageous modifications,
whereat we may congratulate ourselves.

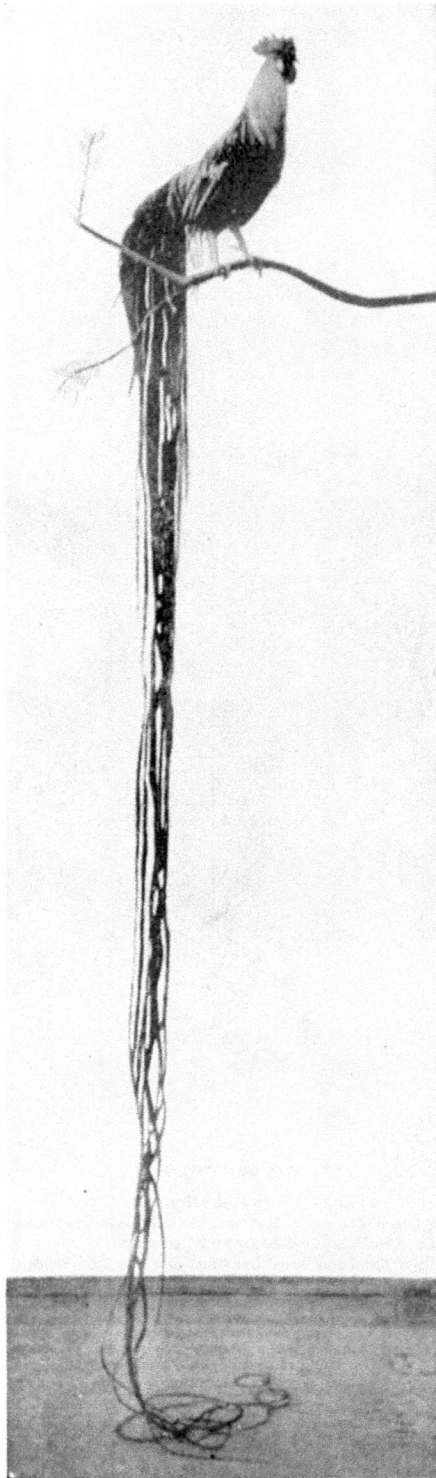

JAPANESE LONG-TAILED FOWL OR TOSA FOWL

In this extraordinary breed, which is believed to be of very ancient origin, the feathers of the tail show continuous growth, reaching 7 to 8 feet, and in extreme cases 18 feet. This seems to be a physiological mutation. The offspring of a cross between a Tosa cock and a white cochin Bantam hen yielded males with the Tosa coloration except that every feather was barred with white. The males had abnormally long middle tail-feathers, but not so long as in the Tosa cock. The female offspring were like Tosa hens.

COMBS OF FOWLS

A. Single serrated comb, as in Leghorns and Minorcas.

B. Pea comb, with three well-marked ridges of little papillæ, the median one a little higher than the others, as in Indian game-fowls and Brahmas.

C. Rose comb, with a flattened area bearing papillæ, and behind these a pike, as in Hamburgs and Rose-combed Dorkings.

The pea character shows definite dominance, thus pea × single yields pea. The rose character also shows definite dominance, thus rose × single yields rose.

But rose × pea yields out of sixteen cases an average of nine "walnuts," a different kind of comb altogether. The walnut comb has no distinct papillæ like the rose, or ridges like the pea. It shows a corrugated surface suggesting a walnut, and there is generally a curious band of bristles crossing the comb at the beginning of the posterior third. The rest of the members of an average sixteen series from rose × pea are three "rose," three "pea," and one single—a result which admits of reasonable Mendelian interpretation.

§ 4

Origin of Variations

Darwin had no theory of the origin of variations, and we must join with him in saying "our ignorance of the laws of variation is profound." This is the central problem of evolution— the origin of the new. Yet certain possibilities have become clearer since Darwin's day. When a white blackbird is hatched, when an albino child is born, when a calf appears without horns or a kitten without a tail, we interpret these variations as due to the dropping out of the relevant hereditary item in the inheritance, and we know that in the history of the germ-cells there are definite opportunities for such losses.

When, on the other hand, an offspring has more than usual of a certain character we can interpret this as due to its getting a double dose—from both sides of the house—of the hereditary item in question. If both parents are very dark and come of very dark stocks, the offspring may be darker still, and the same holds terribly true of a double dose of some disadvantageous character, such as deaf-mutism. The individual life always begins in the fertilised egg-cell, and there may be accentuation of a character, we say, if it is strongly represented both in the paternal and in the maternal hereditary contributions. In the sperm-cell as in the egg-cell there is a complete set of hereditary "factors" or initiatives, and these two sets come into intimate and orderly union in fertilisation. When the fertilised egg develops into an embryo and into a young creature, there may be an expression of some paternal peculiarities and some maternal peculiarities, with a new pattern as the result. It must be understood that although there is a complete assortment of hereditary qualities in the egg-cell and also in the sperm-cell, it is usually only one set that finds expression in the offspring in regard to any particular structure. The child may have its mother's hair, its father's chin. In some cases a father's character as regards some particular feature is seen only in his sons, not in his daughters. But the feature may appear in his daughter's sons.

When the human variant shows a new pattern of a particularly happy kind, we call it "genius"; when the outcome is more dubious we say "crank." And the animal kingdom is full of geniuses and cranks. Our point, however, is just this, that fertilisation offers an opportunity for new permutations and combinations. If we may compare an inheritance to a pack of cards, each hereditary constituent or "factor" corresponding to a card, then there is in fertilisation a re-shuffling, just as there is in the maturation of the germ-cells an opportunity for cards being lost. We may say, then, that an increased knowledge of the history of the germ-cells since Darwin's day has made it possible to understand how certain kinds of variations may arise.

If we probe a little deeper, we see the possibility that the stimuli of outside changes, e.g. of climate, may saturate through the organism *and provoke the complex germ-cells to change.* Thus Professor W. L. Tower subjected potato-beetles at a certain stage of their development to very unusual conditions of temperature and humidity. The beetles themselves were not changed, for these hard-shelled creatures do not lend themselves to external modification. But in a number of cases the *offspring* of the beetles showed remarkable changes, e.g. in colour and markings. And the offspring of these variants did not revert to the grandparental type. In such a case it looks as if an environmental stimulus penetrating through the body serves as the liberator or stimulus of variability in the germ-cells.

It may seem for a moment that this case of the potato-beetles indicates the inheritance of the results of environmental influence. But it must be carefully noticed that the parent beetles showed no modification or acquired character. What happened was that a peculiarity of environment saturated through the body, and started a germinal peculiarity, which all biologists are agreed in regarding as heritable. Similarly, persistent alco-

holism on the part of a strong parent may prejudice the offspring by provoking disturbance in the germ-cells. But this is
very different from the transmission of hardened liver or any
other specific modification. Everyone knows that alcoholism of
parents does not make for vigorous progeny, but it must be
insisted that this does not bear very directly on the technical
problem of the transmission of modification. In most cases
what is inherited in the alcoholic lineage (rarely a long one) is
a constitutional defect, e.g. lack of control. In some cases the
parental intemperance affects the germ-cells prejudicially; though
in some animals the results of experiments do not corroborate
this. It seems to vary with the organism. Finally, the offspring
of an alcoholic mother may be badly handicapped before birth,
but this has as little bearing on the transmission of acquired
characters as the fact that whisky babies do not thrive. It is
not legitimate to re-define "acquired characters"; the term means
—modifications of structure acquired in the individual lifetime
as the direct result of peculiarities in surroundings, food, and
function.

Professor Weismann laid emphasis on the somewhat subtle
idea that the complex germ-plasm, which somehow contains the
whole inheritance, might be prompted to vary by fluctuations in
the nutritive stream of the body. Just as poisons in the blood
may deteriorate the germ-cells in definite ways, so the gentler
influence of slight changes in nutrition may induce the germ-cell
to internal rearrangements which are by and by expressed as
profitable variations. It should not be forgotten that differences
in diet determine whether the grub of a bee is to develop into
a worker or into a queen.

It seems fair to say that the problem of the origin of variations is not so dark as it was in Darwin's time. At the same
time no one can pretend to understand the emergence of the
distinctively new. The germ-cell is a living creature in a single-
cell phase of being, and it may be that its variations are the out-

comes of a primary quality of living creatures, inherent in the germ-cell—the capacity of making experiments in self-expression.

§ 5

As regards Heredity

Darwin was one of the first to show that the mysterious problems of heredity could be attacked scientifically, and his cousin Sir Francis Galton went much further. But it is unfortunate that neither of them knew anything about the Abbé Mendel, who published papers in 1865 which have revolutionised the whole subject. His work remained practically unknown till 1900.

Mendelism

There are three fundamental ideas in Mendelism. The *first* is the idea of "unit-characters," and this requires a little patience. By an inheritance is meant what the living creature is or has to start with, when it is represented by a fertilised egg-cell. Now it has been discovered that an inheritance is, in part, built up of numerous, more or less clear-cut, crisply defined, non-blending characters, which are continued in some of the descendants as definite wholes, neither merging nor dividing. We may think of the colour of the eye, the quality of the hair, the shape of the nose. Strictly speaking, what lies in the inheritance is not the character as seen in the adult but a germinal representative (technically called a "factor" or "gene") of the character. The full-grown character, say the shape of the nose, is, as it were, *a product of the germinal representative and the surrounding influences which operate during development.* It is also necessary to understand that an adult character, like the quality of the hair, may be represented in the germ-cell by several factors. Moreover, one germinal factor, e.g. the initiative for developing dark pigment, may influence several characters in the adult.

HEREDITY IN WILLOWS

(*After Wiesner*)

A, a broad-leaved variety, the one parent.

C, a narrow-leaved variety, the other parent.

B, the hybrid offspring, intermediate between the two. This has an appearance of *blending*, but it may be a case of imperfect dominance, as in the Andalusian fowls. Or it may be that the shape of the leaf depends upon a number of Mendelian unit characters which are not linked together but produce an appearance of blending by their fortuitous distribution in the offspring. If some come from the one parent and some from the other they may neutralise one another, with an *apparent* " blend " as the result.

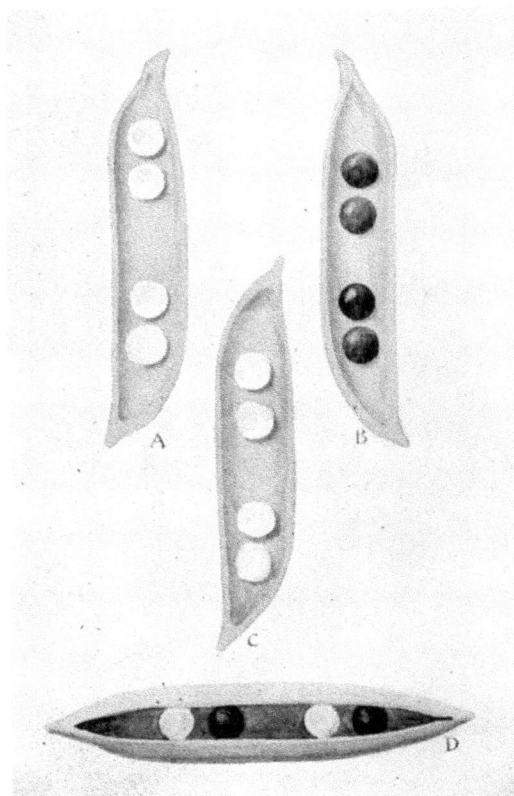

MENDEL'S LAW ILLUSTRATED IN PEAS

A. Pod of a yellow-seeded pea, the one parent (dominant as to seed-colour).

B. Pod of a green-seeded pea, the other parent (recessive as to seed-colour).

C. Pod of the hybrid offspring (the first filial generation), with only yellow seeds. Yellow-seededness is dominant and green-seededness recessive.

D. The next generation (the second filial generation) shows the occurrence of both yellow seeds (left light) and green seeds (shaded dark).

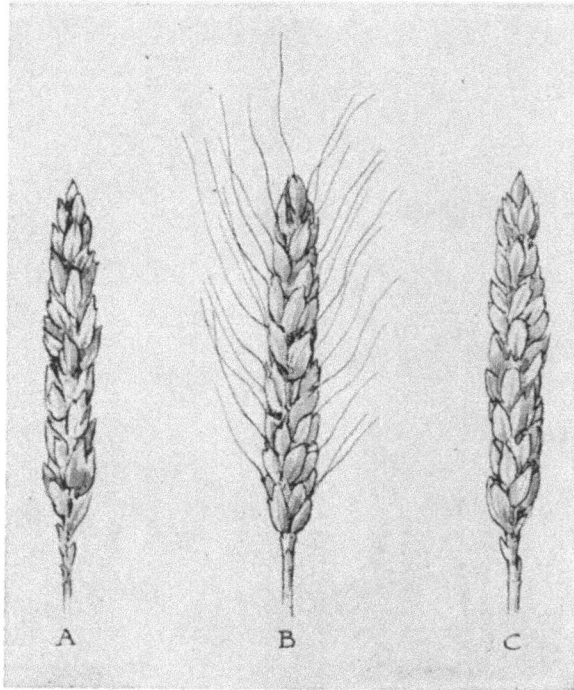

MENDELIAN INHERITANCE IN WHEAT

(After R. H. Biffen.)

A. Stand-up wheat, with no beard, the one parent.
B. Bearded wheat, the other parent.
C. The hybrid offspring, with no beard.

This shows that the beardless condition is dominant and the bearded condition recessive.

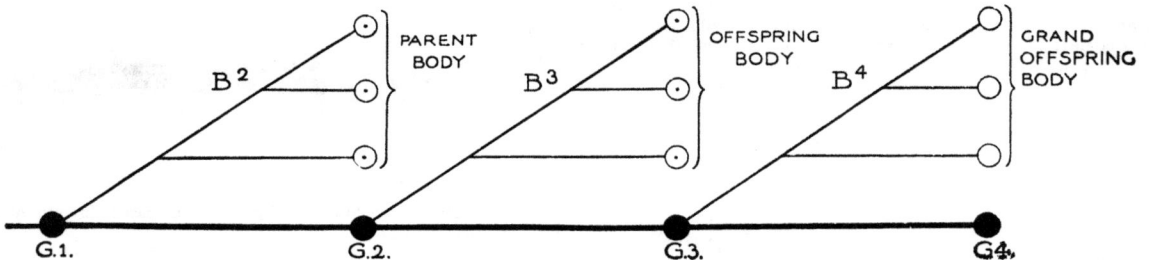

THE IDEA OF GERMINAL CONTINUITY

G1. A fertilised ovum developing into a lineage of body-cells (B²) and a lineage of germ-cells (the dark thick base-line). G2. Germ-cell which starts the offspring of the next generation, with its body-cells (B³) and its germ-cells along the base-line. G3. A germ-cell starting the grand-offspring, with its body-cells (B⁴), e.g. ectoderm, mesoderm, and endoderm, and likewise its germ-cells along the basal "germ-track," G3 to G4.

The base-line represents the lineage or chain of germ-cells; B², B³, B⁴ are the bodies of three successive generations which fall off from the chain. The fundamental idea is that a fertilised egg-cell gives rise to a body *and* the germ-cells of that body.

If a man has his fingers all thumbs, i.e. with two joints instead of three, this peculiarity (called "brachydactylism") is sure to be continued in a certain proportion of his descendants; and we call it a "unit-character." The persistence of the Hapsburg lip in the Royal Houses of Austria and Spain is a good instance of how a unit-character comes to stay for many generations. Night-blindness, or the inability to see in dim light, has been traced through a lineage since near the beginning of the seventeenth century—another illustration of the persistence of a unit-character. We do not precisely know what the germinal factors of the unit-characters are like, but in some cases it is known that they lie in linear order in the nuclear rods or chromosomes. In some instances (though it is impossible in a few words to explain *how*) we know what region of the chromosome the factor occupies. But the most important point is that the unit-characters (or their factors) behave as if they were definite entities, like the radicals in chemistry, which can be shuffled about and distributed to the offspring in some degree independent of one another. Thus in the lineage of the "nightblind" it was not every individual that showed the peculiarity, but only a certain proportion in each generation.

In his masterly book on *Mendelism* Professor R. C. Punnett refers to a unit-character as follows: "Unit-characters are represented by definite factors in the gamete [or germ-cell], which, in the process of heredity, behave as indivisible entities, and are distributed according to a definite scheme. The factor for this or that unit-character is either present in the gamete or it is not present. It must be there in its entirety or be completely absent."

§ 6

The *second* fundamental idea in Mendelism is that of *dominance*. When Mendel crossed a pure-bred tall pea with a pure-bred dwarf pea the offspring were all tall. So he called the quality of tallness dominant to the recessive quality of dwarfness, which the hybrid offspring kept, as it were, up their sleeve. The dwarfness is not expressed in the hybrid peas, but it must be part of the inheritance, for it reappears in a quarter of the progeny of the hybrids if these are inbred or allowed to self-fertilise.

The Japanese have reared a race of peculiar waltzing mice, which have many strange habits, e.g. of dancing round and round. If a Japanese waltzing mouse is crossed with a normal mouse, all the hybrid offspring are normal, the waltzing peculiarity being recessive to normality. But if these hybrid mice are paired together, some of their progeny are waltzers—in the proportion of one waltzer to three normals, which is called the Mendelian ratio. If one of the waltzers of the second generation pairs with another waltzer, the progeny are all waltzers, which shows that the factor for normal locomotion has disappeared from the inheritance along this line. It is a curious fact that one of these second generation waltzers might be conscientiously sold in the market as a pure waltzer, although its parents were normal and one of its grandparents likewise. To return to the beginning, if a waltzing mouse is crossed with a normal mouse, all the offspring will be normal. Normality is dominant; waltzing is recessive. If these normal hybrids pair, their offspring will be 25 per cent. pure waltzers and 75 per cent. apparently normal mice. But of the 75 per cent. apparently normal a third will be pure normals, yielding nothing but normals when bred with others like themselves. But the other two-thirds, though apparently normal, have, like their immediate parents, the waltzing character up their sleeve, for when they are paired together they yield 25 per cent. pure normals, 50 per cent. apparent normals, and 25 per cent. pure waltzers. It is impossible to keep this clearly in mind without some schematic formulation, such as the above.

In the case of the mice the character of normal locomotion

is dominant over the recessive character of waltzing, but it must not be supposd that the dominant character is necessarily the one nearest the normal type. Thus a short tail in cats is dominant (somewhat imperfectly) to the ordinary tail; the appearance of extra toes in poultry is dominant to the presence of the normal four toes; hornlessness in cattle is dominant to the presence of horns.

Among the many characters which are now known to exhibit Mendelian inheritance, the following may be cited, the dominant condition being named first in each case: Normal hair and long Angora hair in rabbits and guinea-pigs; kinky hair and straight hair in man; crest and no crest in poultry; bandless shell in the wood-snail and banded shell; yellow cotyledons in peas and green ones; round seeds in peas and wrinkled forms; absence of awn in wheat and its presence; susceptibility to "rust" in wheat and immunity to this disease; two-rowed ears of barley and six-rowed ears; markedly toothed margin in nettle leaves and a slightly toothed margin. Why one character should be dominant and another recessive is not known; a positive feature, like a banded shell in the snail, may be recessive; and a negative feature, like hornlessness in cattle, may be dominant.

It should be noted that in many cases of Mendelian inheritance the dominance in the offspring is not complete; thus, if black Andalusian fowls be crossed with white ones the progeny are "blue" Andalusians—a sort of diluted black. These "blue" Andalusians do not breed true; when paired together they yield 50 per cent. "blues," 25 per cent. blacks, and 25 per cent. peculiar whites splashed with grey.

§ 7

The *third* fundamental idea in Mendelism is perhaps more difficult to grasp than the others. Mendel supposed that the hybrid between the tall pea and the dwarf pea produced two kinds of germ-cells in approximately equal numbers—one contingent carrying the factor for tallness and the other contingent carrying the factor for dwarfness. In other words, each germ-cell is "pure" with respect to the factor of any particular unit-character. Suppose a long-haired rabbit crossed by a short-haired rabbit, the offspring will be all short-haired. But out of eight ova produced by a female hybrid offspring, four will have the factor for long hair and four the factor for short hair. Similarly, out of eight sperm-cells produced by a male hybrid offspring, four will have the factor for long hair and four the factor for short hair. Suppose these hybrids interbreed, and the fertilisation of the ova by the spermatozoa is fortuitous, then two egg-cells with the short-hair factor will be fertilised by two sperm-cells with the short-hair factor, yielding two quite pure short-haired offspring; two egg-cells with the long-hair factor will be fertilised by two sperm-cells with the long-hair factor, yielding two quite pure long-haired offspring; two egg-cells with the short-hair factor will be fertilised by two sperm-cells with the long-hair factor, yielding two impure short-haired offspring like the hybrid parents; and, finally, two egg-cells with the long-hair factor will be fertilised by two sperm-cells with the short-hair factor, yielding other two impure short-haired offspring like the hybrid parents. So the result must be two pure short-haired offspring, plus four impure short-haired offspring, plus two pure long-haired offspring. If the impure short-haired rabbits are interbred, their offspring (the third filial generation) will show the same ratio, 1 : 2 : 1, more and more exactly the larger the numbers dealt with.

§ 8

Germinal Continuity

One of the great post-Darwinian advances is the recognition of the fact of germinal continuity—made clear by Galton and Weismann. While most of the material of the fertilised ovum is used to build up the body of the offspring, undergoing in a

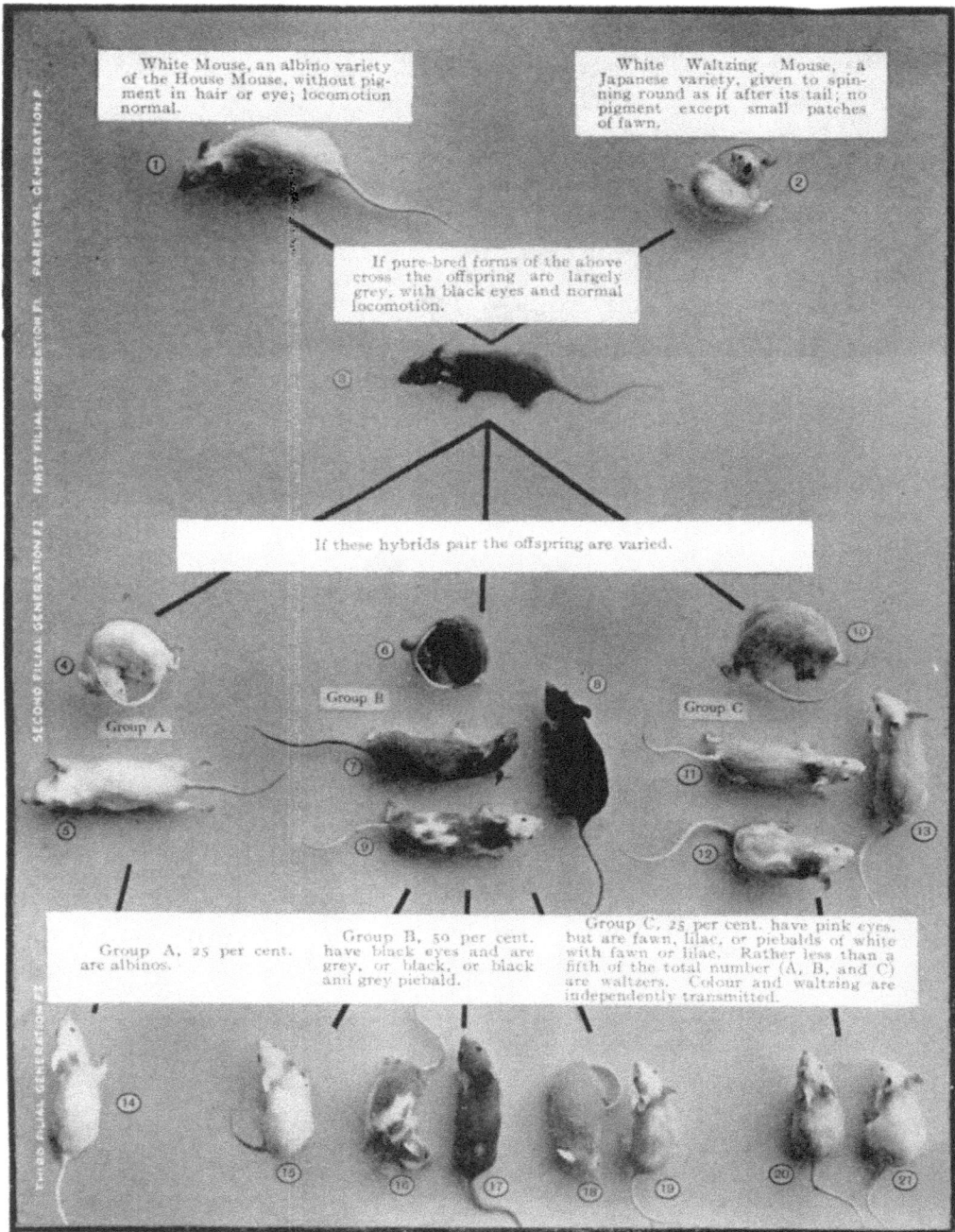

White Mouse, an albino variety of the House Mouse, without pigment in hair or eye; locomotion normal.

White Waltzing Mouse, a Japanese variety, given to spinning round as if after its tail; no pigment except small patches of fawn.

If pure-bred forms of the above cross the offspring are largely grey, with black eyes and normal locomotion.

If these hybrids pair the offspring are varied.

Group A

Group B

Group C

Group A, 25 per cent. are albinos.

Group B, 50 per cent. have black eyes and are grey, or black, or black and grey piebald.

Group C, 25 per cent. have pink eyes, but are fawn, lilac, or piebalds of white with fawn or lilac. Rather less than a fifth of the total number (A, B, and C) are waltzers. Colour and waltzing are independently transmitted.

Members of Group A mated together produce only albinos like themselves.

Members of Group B mated together produce greys only like (3), or a mixed litter of albinos, greys, fawns, and piebalds of these.

Members of Group C mated together produce fawns, lilacs, piebalds of these, and an occasional albino.

Photo: British Museum (Natural History).

MENDELISM IN MICE

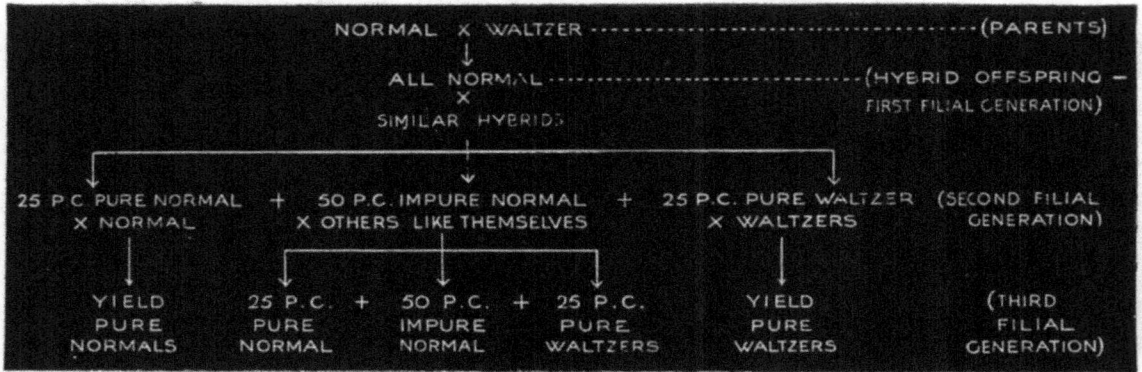

MENDELISM IN MICE

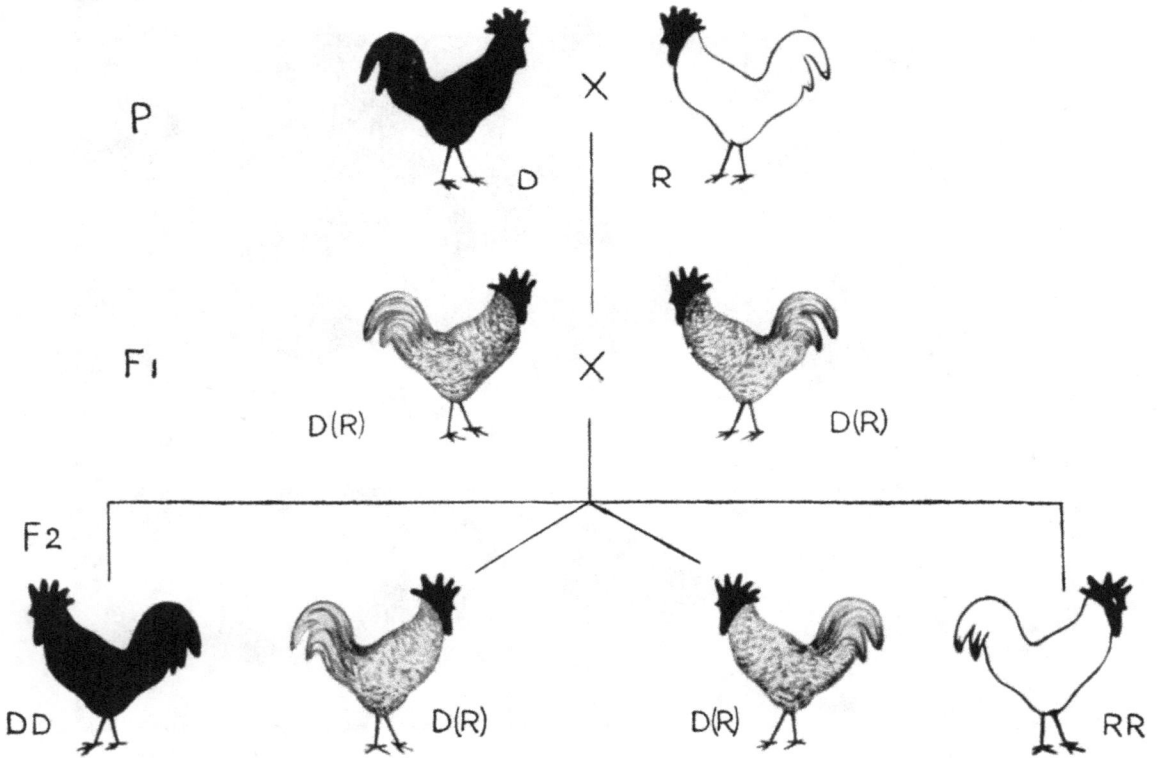

MENDELIAN INHERITANCE IN ANDALUSIAN FOWLS

(*After Darbishire.*)

P, the parents, black (dominant) and white (recessive).

F1, the hybrid generation, "blue" Andalusians, illustrating imperfect dominance.

F2, the second filial generation: 25 per cent. pure blacks ("extracted pure dominants"), DD; 50 per cent. "blues" (impure dominants), D (R); and 25 per cent. whites (extracted recessives), with occasional black spots (RR).

very puzzling way differentiation into nerve and muscle, blood and bone, a residue is kept intact and unspecialised to form the beginning of the reproductive organs of the offspring, whence will be launched in due course another organism on a similar voyage of life. The reproductive cells of any organism are the outcome of embryonic cells which did not share in the upbuilding of that organism, but continued the germinal tradition unaltered. This is suggested clearly in a diagram slightly modified from one devised by Professor E. B. Wilson. Thus the parent is rather the trustee of the germ-plasm than the producer of the child. In a new sense the child is a chip of the old block. The old question was: Does the hen make the egg, or the egg the hen? The modern answer is that the fertilised egg makes the hen *and the eggs thereof*. The fact of germinal continuity explains the inertia of the main mass of the inheritance, which is carried on with little change from generation to generation. Similar material to start with; similar conditions in which to develop; *therefore* like tends to beget like. As Professor Bergson puts it, "life is like a current passing *from germ to germ* through the medium of a developed organism."

§ 9

As regards Selection

When we are interpreting the past history of animals, we utilise factors which are seen in operation to-day, just as the geologist does when he is interpreting scenery. It is satisfactory, therefore, that post-Darwinian investigations have *demonstrated* some modern instances of selection *at work*. Let us take a simple case. The Italian naturalist Cesnola tethered some green Mantises with silk thread on green herbage, and found that they escaped the eyes of birds. Similarly, when the brown variety was tethered on withered herbage. But green Mantises on brown herbage and brown Mantises on green herbage were soon picked off. Discriminate selection was at work.

When we are concerned with making a good lawn we may pursue two methods. We may eliminate the weeds or we may foster by suitable tonics the growth of the grass. Similarly, in Nature's sifting there is *lethal selection,* which works by eliminating the relatively less fit to given conditions of life, and there is *reproductive* selection, which works through the predominant increase of the more successful. Darwin never thought simply of Natural Selection; he always emphasised its manifold and subtle modes of operation. He saw, for instance, what some of his successors missed, that the sifting need not in the least involve a sudden cutting off of the relatively less fit, for a shortened life and a less successful family will in the long run bring about the same result as a drastic pruning. It should not be necessary to point out that "the survival of the fittest" does not *necessarily* mean the survival of the strongest or cleverest or best; it simply means "fittest" relatively to particular conditions. The tapeworm is a fit survivor as well as the Golden Eagle.

Darwin realised what some of his successors have missed, that even slight peculiarities may be of critical moment when tested in relation to the complex web of life in which the creature has its being. This is very important in regard to the general progressiveness of evolution—that new departures are sifted in reference to a slowly wrought out and firmly established system of inter-relations. (See the article on Inter-relations.)

§ 10

Sexual Selection

Many male animals, such as stags, antelopes, sea-lions, black-cock, and spiders, fight with one another at the mating time, competing for the possession of females. According to Darwin, "the strongest and, with some species, the best-armed of the males drive away the weaker; and the former would then unite with the more vigorous and better-nourished females, because they are the first to breed. Such vigorous pairs would surely rear a larger number of offspring than the retarded females, which would be compelled to unite with the conquered and less powerful males, supposing the sexes to be numerically equal; and this is all that is wanted to add, in the course of successive generations, to the size, strength, and courage of the males, or to improve their weapons" (*Descent of Man,* 2nd ed., p. 329). Similarly, there would be a premium on those male characters that are useful in the recognition and capture of the females, e.g. large olfactory feelers in moths and strong claspers in skates.

The term "sexual selection" was used by Darwin to include all forms of sifting in connection with mating, but prominent among these was the preferential behaviour of the female. "Just as man can give beauty, according to his standard of taste, to his male poultry . . . so it appears that female birds in a state of nature have, by a long selection of the more attractive males, added to their beauty or other attractive qualities." In the courtship, which is often elaborate, the female selects—in a literal sense.

Darwin was well aware of difficulties besetting his theory of sexual selection, and his fellow-worker Alfred Russel Wallace was one of his severest critics. There has to be proof that some of the males are actually disqualified and left out in the cold. But Darwin indicated that the sifting would work even if the less successful males were not entirely eliminated. Moreover, in some cases the female's preference goes to great lengths; thus a female spider often kills a suitor who does not please her.

It is difficult, again, to prove actual "choice" on the female's part. But there are undoubted cases of preferential mating, whatever the psychology of the process may be. Some critics, like Wallace, have pointed to the difficulty of crediting the female with a capacity for appreciating slight differences in the decorativeness, agility, or musical talent of her suitors. But the modern answer is simply that the accepted mate is the one that most strongly evokes the pairing instinct, and that it is not necessary to credit the female with any analytic weighing of the merits of the various males. The details must count, if there is anything in the theory, but they may count, not as such, but as contributing to a general impression of interesting attractiveness.

To point out that certain masculine features, such as antlers, are congruent with the male constitution, just as certain feminine features, such as functional milk-glands, are congruent with the female constitution, is getting behind the question of selection to that of the origin of the variations which form the raw materials of the sifting process—an interesting line of inquiry which has been followed by Geddes and Thomson in their *Evolution of Sex.*

Another important consideration arises when we think of the frequent intricacy and subtlety of the courtship habits (see Pycraft's *Courtship of Animals*). There must be some deep racial justification for this. Groos has suggested that the female's coyness is an important check to the male's passion, which tends to be too violent. Julian Huxley has suggested from his fine study of the Crested Grebe that the courtship ceremonies establish emotional bonds which keep the two birds of a pair together and constant to each other.

§ 11

CONCLUSIONS

1. If Darwinism means the general idea of evolution or transformism—that higher forms are descended from lower—then it stands to-day more firmly than ever.

2. If Darwinism means the particular statement of the factors in evolution which is expounded in *The Origin of Species, The Descent of Man,* and *The Variation of Animals and Plants under Domestication,* then it must be said that while the main

HALF-LOP RABBIT

A half-lop rabbit, after Darwin, an instance of a variation which seems to be rather uncertain in its inheritance. The peculiarity is that one of the ears hangs down, whereas in "full-lops" both ears do. The pendent ear is often broader and longer than the upright one, an unusual asymmetry. Darwin noted that when the half-lopped condition occurs, whether in one parent only or in both, there is nearly as good a chance of the progeny having both ears full-lop as if both parents had been full-lopped.

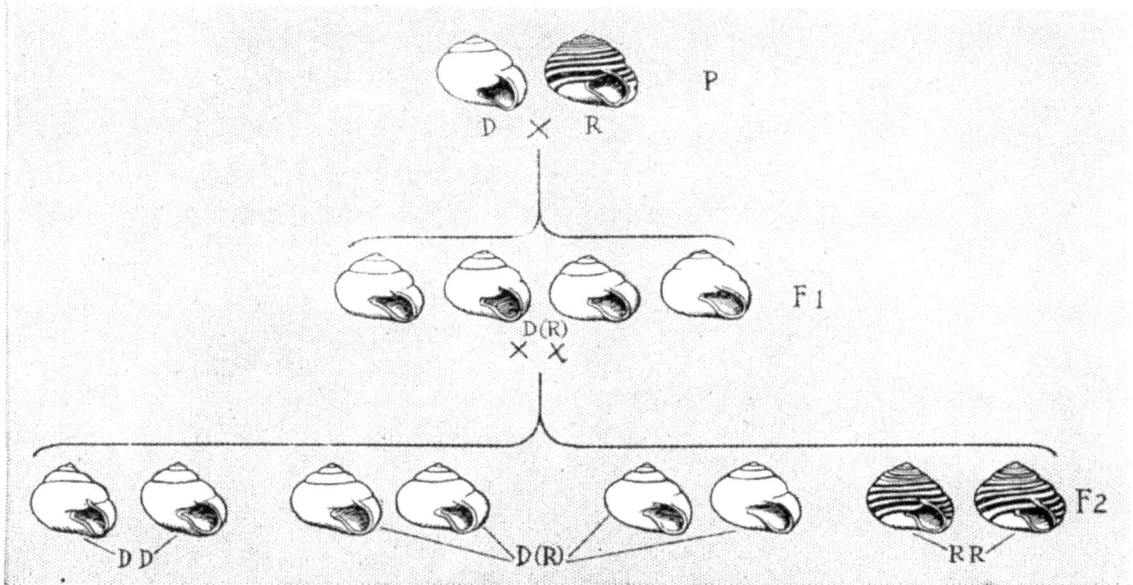

INHERITANCE IN SNAILS, WITH BANDLESS AND BANDED VARIETIES
(After Lang.)

When bandless Wood-Snails (*Helix nemoralis*) or bandless Garden Snails (*Helix hortensis*) are crossed with banded individuals of these species, each will make a nest in the ground and deposit half a hundred eggs or more. A snail is always hermaphrodite, producing eggs and sperms; but the eggs of one snail (banded, let us say) are fertilised by the sperms of another snail (bandless, let us say). Let us follow the eggs of a banded individual, fertilised by the sperms of a bandless individual snail. They will develop into individuals whose shells are all bandless D(R). The negative quality "bandless" (D) is dominant; the positive quality "banded" (R) is recessive. If the bandless hybrids (F1) pair together, the offspring (F2) will be: 25 per cent. pure bandless—extracted dominants (DD); 50 per cent. impure dominants D (R), in appearance bandless; and 25 per cent. pure banded—extracted recessives (RR). If we had started with the eggs of a bandless individual, fertilised by the sperms of a banded individual, the result would have been the same.

THE RUFF (*Macheles Pugnax*)

(*From specimens.*)

The Ruff is a polygamous bird of the plover family. It used to nest abundantly in Britain, but is now hardly more than a bird of passage. In winter the two sexes—ruffs and reeves—are very similar in plumage, but the females are smaller. At the breeding season in spring there is very marked sex-dimorphism. The face of the male becomes covered with little yellow warts, the head is adorned with erectile tufts of feathers, and the fore-neck develops a large "ruff" of feathers which can be raised and depressed according to the state of excitement. In the tufts and the "ruff" there is extraordinary variability of colouring, e.g. white, rufous, or black, with or without bars. It is said that no two are alike. This illustrates the occurrence of countless small variations. The males assemble in little parties at the breeding-time and fight with great vigour hour after hour, but without doing much damage. According to Darwin's theory of sexual selection the tournaments have been factors in the evolution of the sex-dimorphism.

ideas remain valid there has been development all along the line. Darwinism has evolved, as every sound theory should.

3. In regard to the raw materials of evolution, there is greater clearness than in Darwin's time as to the contrast between intrinsic variations of germinal origin and bodily modifications imprinted from without, and there are grave reasons for doubting whether the latter do as such affect the race at all. There is still to be heard the slogan, "Back to Lamarck!" but there can be no return to any crude Lamarckism. If the individual gains and loses, the individual indents and prunings, really count as such in racial evolution, it must be in some subtler way than is suggested by the giraffe getting its long neck by ages of stretching, or the deep-sea fish becoming blind by generations of darkness and disuse. There should be no haste to close any door of reasonable interpretation, still less of experimental inquiry, but there is at present amongst zoologists widespread agreement with Sir Ray Lankester's pronouncement that one of the notable advances since Darwin's day has been getting rid of the Lamarckian theory of the transmission of individually acquired characters or imprinted bodily modifications. Of course, counting of heads is no argument; but the facts are not at present in favor of the Lamarckian view. But we may perhaps look for an evolution of Lamarckism as well as of Darwinism!

4. Darwin based his theory of evolution very deliberately on the fluctuating variations which are always occurring. Given time enough and a consistent sieve (the struggle for existence), will not Nature achieve more or less automatically what man reaches purposefully in his breeding of cattle and cultivating of wheat? But modern Darwinism, while holding fast to this, welcomes the demonstration that brusque discontinuous variations or mutations are common, and that they are very heritable. All of a sudden, it appears, the sporting Evening Primrose may produce an offspring which is potentially a new species.

5. Darwin meant by "fortuitous variations" that he could

not give any formula for the causes of the novelties he observed. No doubt he also meant that the organism in varying was not aiming at anything. And yet he laid great stress on what he called "the principle of correlated variability"—an idea of great importance—that when one part varies other parts vary with it, "being members one of another" as St. Paul said. In other words, a particular germinal change may have a number of different outcrops or expressions. But the more correlation there is, the less reasonable will it be to speak of fortuitousness. And one of the changes since Darwin's day is the recognition that variations are often very definite—just as they are among crystals.

6. Another change from Darwin is the Mendelian idea of unit-characters, which behave like entities in inheritance. They are handed on with a strong measure of intactness to a certain proportion of the offspring. Their "factors" in the germ-cells are either there or not there. Sometimes, at least, these unit-characters arise as mutations, and thus we have an answer to Darwin's difficulty that abrupt changes would be averaged off in intercrossing. Unit-characters do not blend.

7. Since Darwin's day there has been, in a few cases, definite proof of natural selection at work; the different forms of selection have been more clearly disentangled; the subtlety of Darwin's idea of selection has been confirmed; the reality and the efficacy of preferential mating has been much criticised, but Darwin's theory of sexual selection has in its essentials weathered the storm. In proportion as new departures come about suddenly by brusque mutation, the burden to be laid on the shoulders of selection will be lessened. In so far as the selection is in relation to a previously established system of inter-relations, there will be a reduction of the fortuitous in the process; and the same will be true in proportion to the degree in which the organism takes an active share in its own evolution—as it often does.

8. Modern biologists are inclined to put more emphasis on

"Isolation" than Darwin did, meaning by "Isolation" all the ways in which the range of intercrossing is restricted and close in-breeding brought about.

When we use the term Darwinism to mean, not his very words, but the living doctrine legitimately developed from his central ideas of variation, selection, and heredity, we may say that Darwinism stands to-day more firmly than ever. It has changed and is changing, but it is not crumbling away. It is evolving progressively.

This is only an "outline" of a great subject, and it is not an article that he who runs can read. It is very important to avoid dogmatism in regard to an inquiry which is still relatively young. There was not much scientific evolutionism before Darwin's day. The writer has not concealed his opinion in regard to such a question as the transmission of acquired characters, but it is not suggested that this is the only possible opinion. It may be recommended that readers to whom the subject is comparatively new, and to whom it appears full of uncertainties, should write out their ideas in a definite way and then compare them carefully with the relevant paragraphs in the article. It is all too easy to go off on a wrong tack, and this should be guarded against by patient study. For the problems of evolution are fundamental.

BIBLIOGRAPHY

The classic works of DARWIN, WALLACE, AND HUXLEY.

BUTLER, *Evolution Old and New* (1878).

CLODD, *Story of Creation: a plain account of Evolution* (1888).

CONKLIN, *Heredity and Evironment in the Development of Men* (1915).

CONN, *Evolution of To-Day* (1886).

CRAMPTON, *The Doctrine of Evolution* (1911).

DENDY, *Outlines of Evolutionary Biology* (1912).

GEDDES AND THOMSON, *Evolution* (Home University Library, 1911).

HAECKEL, *Evolution of Man.*

KELLOGG, *Darwinism To-day* (1907).

LULL, *Organic Evolution* (1917).

McCABE, *A B C of Evolution* (1920).

METCALF, *Outline of the Theory of Organic Evolution.*

PUNNETT, *Mendelism* (1919).

SCOTT, *The Theory of Evolution* (1917).

SEWARD (Editor), *Darwin and Modern Science* (1909).

THOMSON, *Darwinism and Human Life* (1910); *Heredity* (1919); *The System of Animate Nature* (1920).

WALLACE, *Darwinism* (1889).

WEISMANN, *The Evolution Theory* (1904).

XII

NATURAL HISTORY

NATURAL HISTORY

I. BIRDS

IN previous chapters of this book we have discussed the evolution of animals in general, the inclined plane of behaviour and the everyday life of the body, and it has been necessary to make many references to birds. But there are good reasons for devoting a special chapter to this great class. Birds have entered closely into human life, and in manifold ways. They supply food, and they are the poet's symbols. Their feathers keep us warm at night, and wing the arrow of the bowman. Birds save the world from the continual menace of prolific insects, and they gave the priests a basis for their auguries. To birds we must trace the enormous nitrate beds of Chili which have fertilised the soil of half the world, and we may thank them too for a share in the impulse that has led man to his mastery of the air. Moreover, most birds are joys for ever. Biologically regarded, birds are of supreme interest in their solution of the problem of flight—so different from that of insects, Pterodactyls, and bats; in their variability and plasticity within a comparatively narrow range; and in their fascinating behaviour with its remarkable blending of instinctive and intelligent activities.

§ 1

Millions of years ago the evolution of birds from a reptilian stock began, as has been already described in an early chapter of this work. At first sight it is not easy to see any resemblance between Birds and Reptiles, the one group warm-blooded, conspicuously active, and gloriously beautiful, the other cold-blooded, often sluggish, but perhaps also beautiful in their way. What kinship can there be between the falcon in the sky and the lizard on the wall? The student of comparative anatomy answers that the evidences of similarity are overwhelming: bone by bone the two creatures are built up on a plan that is certainly to a very great extent the same, however much the final products may be modified and adapted. Without much preliminary study of anatomical structure, these points might be difficult to apprehend and appreciate, and we cannot discuss them here; we must accept the verdict of the experts, and admit that birds are the descendants of a reptilian stock—not necessarily of any present-day group of reptiles, but rather of a common ancestor in the immensely remote past. Just one simple point of similarity between the two groups may be mentioned, the fact that both lay eggs, and eggs which are indeed closely alike in several respects.

The Dawn of Bird-Life

We may imagine the ancestral forms as small lizard-like animals, making the first beginnings of the kind of life which we see to great perfection in the birds of to-day. Real power of flight would at first be absent among these early ancestors, but we may think of it as foreshadowed by a great power of leaping from branch to branch in the trees of the primeval forest, where these far-off ancestors of our birds had taken refuge from their terrestrial enemies. We may picture them as making the most of their arboreal haunt, probably using holes in the tree-trunks in which to hide and to lay their eggs, and gradually developing a greater and greater agility in moving about above ground in search of food, and in escape from such enemies as were still able to molest them.

This mode of life would tend, generation after generation, to produce strong propelling hind-limbs, together with fore-limbs, armed with hook-like claws useful for taking hold at the end of each jump and for more leisurely clambering at other times. The crucial step in the evolution of the true bird stock, however, must have been the acquisition of powers of real flight. At an early stage the fore-limbs would be held out sideways during each leap, and later the surface area would become enlarged by the development of a fold of skin between each of these limbs and the body. Later yet this fold would become still more important, and its area would be still further increased by the transformation of its covering scales into some primitive form of feather. Longer and longer leaps would become possible, from branch to branch and from tree to tree, as these aids to gliding flight improved. Finally, the last great step would be taken when a beginning was made of the active use of the primitive wings to prolong still further, until at last indefinitely, the distances possible by leaping and gliding alone.

It is a curious history, this tale of the origin of birds. In the first place we seem to see the earliest ancestors as a feeble reptilian race driven from the ground and taking refuge among the branches. There followed ages of arboreal life during which the great adaptation of flight originated and was made perfect. Then came a day when the new race of birds, fortified with the great advantage of mastery of the air, spread abroad from the forests— to reconquer the ground-level, to find their bread upon the waters, to cross the seas to distant isles, and to defy the rigours of climate by their ability to "change their season in a night." So to-day we have birds peopling the whole earth and filling every land with the abundant beauty of their plumage and their song, and with the immense wonder of their eager, spirited lives.

§ 2

Flightless Birds

It is a strange side-issue, too, to find that the priceless gift of flight has not always been preserved. Over and over again since the reconquest of the ground-level, there have been birds which have discarded the faculty which was the making of their race; over and over again, also, they have paid the extreme penalty. Sometimes size and strength, sometimes an aquatic life, sometimes an island home, has been the factor giving security in place of flight, but with new conditions the exchange has frequently proved to be unfortunate: too often, in recent cases, the new condition has been the advent of modern Man and his civilisation.

Several flightless species are indeed numbered among the birds which have become extinct within historic times. Among the Maoris of New Zealand there was a traditional knowledge of a giant running bird which they called "Moa," but which they had exterminated before the arrival of white men; from the bones and other remains which have been found in some quantity the birds appear to have been large members of the Ostrich tribe, one species standing 12 feet in height. A related bird of similar history was the Æpyornis of Madagascar, which forms the subject of the delightfully imaginative story by Mr. H. G. Wells. This bird is sometimes identified with the legendary "Roc" of the *Arabian Nights;* not only its remains but also its eggs have been found, and an egg in the British Museum (Natural History) measures more than 13 inches in length and 9½ in breadth.

The Dodo

"Extinct as the Dodo" has become a proverbial expression. The saying refers to a bird allied to the Pigeons, about the size of a Swan, and of clumsy and uncouth appearance. It was quite flightless, and lived in security in Mauritius until the island was visited by Dutch sailors in the sixteenth century. The hogs which these men brought with them were largely responsible for the subsequent rapid extermination of the birds, and now the Dodo is known only from some remains in museums and from the quaint drawings and descriptions of the early voyagers.

MODEL OF THE EXTINCT DODO

The Dodo was a large flightless pigeon inhabiting Mauritius. It became extinct in the sixteenth century soon after the arrival of Dutch sailors on the island: the pigs which these men brought with them played a large part in the extermination.

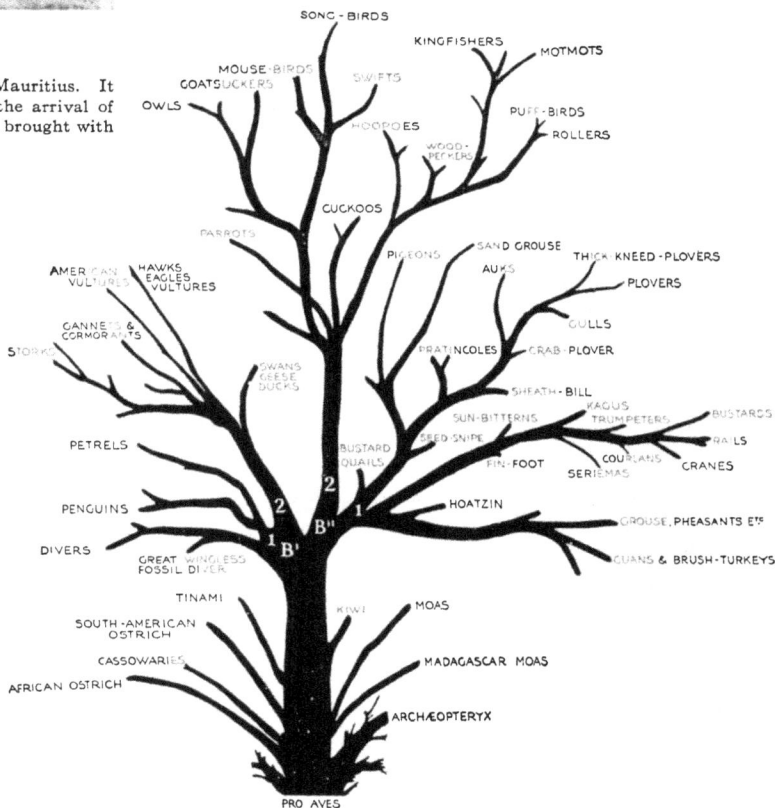

After W. P. Pycraft.

THE EVOLUTION OF BIRDS

A diagrammatic attempt to express the relationship of the main groups of present-day birds, based on a study of structural affinities. The main stock is shown as arising from the ancestral "Pro-Aves," akin to the ancient Reptiles. Offshoots low down represent the extinct Archæopteryx and the different Ostrich-like birds, present and recent. Higher up two main branches (B', B") are shown, each of them dividing again into two (1, 2; 1, 2) and then into the twigs representing the various groups.

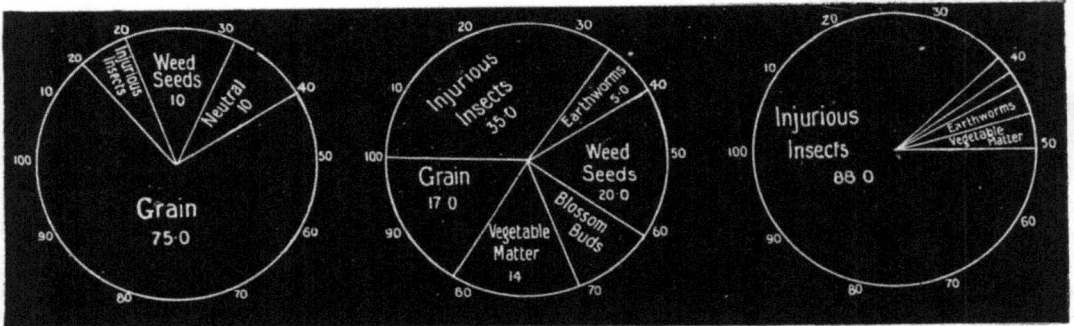

Adult Sparrow in
an agricultural
district.

Adult Sparrow in a
fruit-growing
district.

Nestling
Sparrow

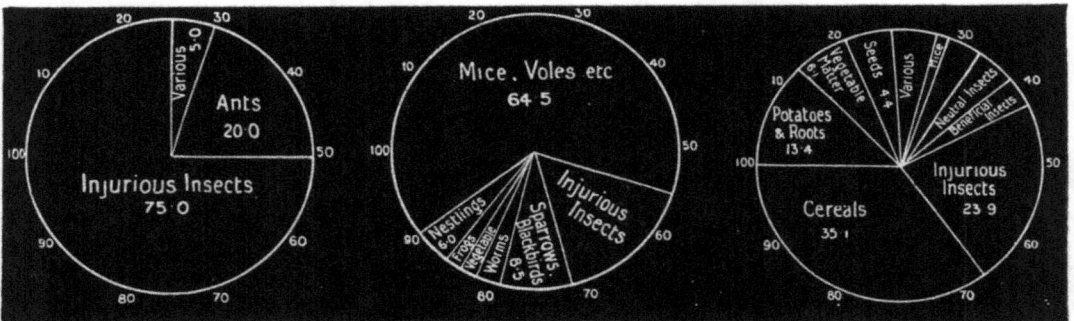

Green Woodpecker.

Kestrel.

Rook.

BIRDS AND THEIR FOOD

This diagram illustrates to what extent certain birds are harmful or useful in respect of our food crops. The evidence is obtained by examining the bird's crop or gizzard at the time of death. The calculations were based on statistics published by Dr. Collinge in the *Journal of the Board of Agriculture*.

The Ostrich Tribe

Among the birds of the present day, the Ostrich tribe and the Penguins are the principal examples of flightlessness. The Ostrich and its kin are for the most part birds of large size, possessing a soft, hair-like plumage, diminutive wings, and strong legs; they are capable of running at great speed across open country, and also of kicking with suddenness and force. Their breastbones lack the pronounced "keel" which is so noticeable in most birds, and which serves for the attachment of the great muscles for working the wings in flight. Best known, of course, is the African Ostrich, now being domesticated by man for the sake of its plumes, but there are also several kinds of American Ostriches or Rheas in South America, and of Cassowaries and Emus in Australasia. Unlike their fellows are the Kiwis of New Zealand, birds of no great size, timid and nocturnal in habit; their long beaks and their hair-like plumage combine to give an exceedingly quaint appearance, and there are no visible wings.

Penguins

The Penguins are rather a different case, for their wings have by no means fallen into disuse; they have become, instead, adapted for swimming. There are many different kinds, but all belong to the Southern Hemisphere, and most of them to the far south. Many Antartic explorers have brought back tales of their life, but it is to Dr. Murray Levick, who was with the *Terra Nova* in 1910, that we owe one of the best accounts, relating particularly to the Adelie Penguin. These flightless birds will return, "over hundreds of miles of trackless sea," to the same "rookeries" year after year to breed. Dr. Levick describes how the first penguin arrived at the "rookery" at Cape Adare towards the middle of October, the southern spring, and how four days later the birds were coming in across the still unbroken sea-ice in such numbers that they formed a line stretching northwards as far as the eye could see; within a month the colony was some three-quarters of a million strong.

The Adelie Penguin builds a large nest of stones, the only material available, and the uses of this are evident when the thaw comes and the ground is covered with water and slush. In this nest two large eggs are laid, and one of the parents goes off to the sea to feed while the other remains to incubate. The bird which leaves may be away for a week or ten days, and the other may therefore not break its fast for as much as four weeks in all.

"I know of no other creature," says Mr. Herbert G. Ponting, "from which man may learn a finer lesson of how resolution and steadfastness may overcome every difficulty than from the Adelie Penguin." Their bravery is amazing; no blizzard, however violent, will drive these birds from their nests in the wild Antarctic regions. Mr. Ponting relates that they are found sitting on their nests buried deep in the snow. Wondering where the birds had disappeared to after a blizzard, he set out to investigate. "As I was struggling about, wondering whether my penguin investigations had come to an abrupt end, I was almost 'scared out of my life' by a muffled squawk, and felt something wriggling under my foot. I had stepped on the back of a sitting penguin—buried nearly two feet deep in the snow. As the victim struggled out, loudly protesting its wrath at this outrage, we were convulsed with laughter; then, roused by our noisy mirth, scores of black heads, with 'gollywog' eyes, suddenly protruded from the snow —to see what all the fuss was about. That was how we discovered them! They had *not* deserted the place; but were attending to their domestic duties *under* the snow—patiently waiting for it to blow away. There were penguins everywhere; it was impossible to walk without stepping on them."

The penguins are fond of all manner of amusements; leaving their young under the protection of a few of the old birds, most of the parents go off to disport themselves on the ice or in the water. "They will string out behind a leader and make for the near ice-floes, the party sometimes porpoising along the water, then toboganing over the ice. They followed in a line behind the leader, doing exactly as he did. The fun became fast and furious, and I suppose they got a bit winded, for after a while the courier gave them a rest. Following his lead they sprang on to an ice-raft; then, still imitating his example, they settled down on their breasts and basked awhile in the sunshine—prior to doing a few more laps. That they all thoroughly enjoyed the game, there could be no possible doubt." [1]

The Emperor Penguin is the largest species and may stand over four feet high. Unlike the Adélie it nests, or rather lays its single egg, on the sea-ice itself, and it is remarkable for breeding in mid-winter.

The Emperor Penguin

Incubation lasts for as much as six or seven weeks, but the task is shared, not only by both parents, but by the strangely large number of barren birds living in the colony. The chick has the rather doubtful advantage of a number of foster-parents all desirous of participating in its care, a strange condition of things which was well described by Dr. A. E. Wilson, who afterwards shared Scott's tragic fate on the return journey from the Pole: "What we actually saw, again and again, was the wild dash made by a dozen adults, each weighing anything up to ninety pounds, to take possession of any chicken that happened to find itself deserted on the ice. It can be compared to nothing better than a football 'scrimmage' in which the first bird to seize the chick is hustled and worried on all sides while it rapidly tries to push the infant between its legs with the help of its pointed beak, shrugging up the loose skin of the abdomen the while to cover it. That no great care is taken to save the chick from injury is obvious from an examination of the dead ones lying on the ice. All had rents and claw-marks in the skin, and we saw this not only in the dead

[1] H. G. Ponting, *The Great White South.*

but in the living. The chicks are fully alive to the inconvenience of being fought for by so many clumsy nurses, and I have seen them not only make the best use of their legs in avoiding such attentions, but remain to starve and freeze in preference to being nursed. Undoubtedly, I think that of the 77 per cent. that die before they shed their down, quite half are killed by kindness."

§ 3

Flying Birds

With this strange and rather terrible picture of the early life of the Emperor Penguin amid the rigours of the Antarctic climate and on the naked ice of the frozen sea, we may turn from flightless to flying birds. The flightless birds, indeed, represent digressions from the main line of descent, and cannot be regarded as stages in the evolution of modern flying birds from the ancient forms which first mastered flight in the forests of long ago.

Birds share with mammals the distinction of being "warm-blooded," that is to say, having a high and constant body temperature independent of surrounding conditions. We may take this as an index of a high degree of vitality and of an advanced position in the evolutionary scale, and we shall indeed find many other features which lead towards the same conclusion. Birds are noteworthy for alertness of mind and body, for quickness of movement, and for their mastery of the air. They have highly developed habits and complex instincts: they are in turn combative, amatory, parental, cunning in pursuit and escape, and in very many cases there is a surpassing beauty of plumage and voice which compels our intense admiration.

"Beast" is one of those words of variable and confused sense which drive men of science to use a language of their own, but the term "bird" scarcely needs to be defined, for its everyday meaning is also scientifically accurate. This fact may perhaps be attributed to the existence of certain very distinctive characteristics

Photo: Royal Scottish Museum.

GREAT AUK (WITH EGG)

A large flightless species related to the Razorbill and the Guillemot. It became extinct about the middle of last century, but was formerly common on the coasts of Iceland and Newfoundland and not unknown in British waters. Specimens and eggs are much prized rarities nowadays, and will command several hundred pounds if in good condition.

Photo: James's Press Agency.

THE KIWI OR APTERYX

A small member of the Ostrich tribe, found in New Zealand. It is not only flightless, like its relatives, but also lacks visible wings. It is a shy bird of nocturnal habits: the beak, used for probing, is long and sensitive; the legs are powerful; the plumage is long and hair-like.

GANNET—A LESSON FOR AVIATORS

THE GANNET—PUTTING ON THE BRAKE

'The instant fish are descried swimming close to the surface, the Gannet comes wheeling round; down go the brakes 'hard all': its depressing planes are tilted; it sweeps downwards, half closing its wings; and then falls like a streak of white light, plumb to the water"—and there is one fish less in the sea.

common to all birds, and to a 'large measure of uniformity in general appearance among the nearly twenty thousand different species which are known to science; there are, it is true, wide differences in size, in coloration, and in manner of life, but there are no gross divergences in form comparable to those found, for instance, among mammals—between the tiger and the goat, the kangaroo and the elephant, or the bat and the whale.

This distinctiveness and this uniformity may both be accounted for in one word—Flight. The whole body of the bird is adapted to this habit of flying. The bird's skeleton is a wonderful study from this point of view, but here it will suffice to mention the external features. Flight has brought with it feathers, and these are a unique feature: all birds have feathers, and nothing that is not a bird possesses any trace of them. Furthermore, the function of flight has secured a virtual monopoly over the fore-limbs, and it has thus brought two other striking adaptations in its train —a bird is of necessity a biped, walking on its two hind-limbs, and its mouth has had to take the place of a hand, thus leading to the evolution of a long flexible neck, and of a hard beak which is often wonderfully adapted to the feeding habits of the particular species.

§ 4

The Flight of Birds

Birds are, of course, true heavier-than-air machines, and in former days man used to strive to learn their secret for the purposes of the flying-machines which his heart desired; but within the last few years the main physical principles of the aeroplane have become so familiar that we may perhaps reverse the process by using them in the description of our present problem! Just as gliders preceded aeroplanes, so gliding flight may, as we have seen, have been the beginning of the mastery of the air in the case of birds; and it is in gliding that the artificial machine and the bird are most alike. In both cases advantage is taken of the resistance of the air, and of the consequent upward tendency imparted to a body moving horizontally and having a flat inclined under-surface.

When we come to active flight a difference is at once obvious: the aeroplane propellers supply a motive force independently of the planes, while in the bird the wings are both propellers and planes at the same time. There is, indeed, a further difference in that the aeroplane's propellers, during level flight at least, exert force purely in a horizontal direction, the lifting force being wholly due, as in gliding, to air resistance. In the bird the wing-strokes themselves supply part of the lifting power, as well as propelling the body forwards. Nor must we forget the bird's tail, which plays a part in steering and balancing as in the case of the aeroplane rudder; it is also often used as a brake, without which many a swiftly pouncing bird of prey would be apt to dash itself to destruction on the ground.

Some of the larger birds are adepts at soaring, and can remain in the air for a long time with motionless wings, and can even rise in slow spiral ascent to a great height. The late Mr. F. W. Headley, a keen and exact student of the flight of birds, came to the conclusion that this feat was inexplicable except on the supposition that advantage was taken of up-currents in the air, the bird's actual motion being merely a gliding one. He pointed out that gulls are adepts at this when flying above the edge of a cliff, but that they cannot do it at sea, where, as aviators and air travellers know, there are not the vertical disturbances caused by the varying ground-level temperature and by the changing elevation of dry land. Another feat, namely hovering, is familiar in the hunting methods of the Kestrel, which maintains a stationary position for an appreciable time. Against a strong wind it would be easy to maintain a *ground* speed of *nil,* and it would be possible even with motionless wings. In still air, however, the ordinary gliding basis of flight is in abeyance, and altitude must be maintained by sheer vertical force of wing-stroke, the bird being thus more nearly equivalent to a helicopter than to an aeroplane.

Speed and Altitude

The aviators of to-day compete to establish records for speed, for endurance, and for altitude. How do birds stand in these respects? As regards speed, in the first place one must remember the difference between "ground speed" and "air speed." Both the aeroplane and the bird can, for a certain expenditure of power, attain a certain velocity in the body of air in which they are, but the velocity as measured from the ground may be a very different thing. Thus an aeroplane travelling at 100 miles per hour in a 20 miles per hour wind may seem from the ground to be going at 120 miles or at 80 miles per hour, accordingly as it flies with or against the air-stream; so also, of course, with the bird. All our speed records of birds, except a few made from aeroplanes, are necessarily in terms of ground speed, and in many cases the particulars necessary for a wind correction are unhappily wanting.

What are some of the actual figures? The available evidence has recently been summarised by Colonel Meinertzhagen, with special reference to speed during migration; he concludes that a bird has an ordinary pace, which is the one used in migratory flight, and an accelerated pace of which it is capable for a short distance under stress of danger or in other special circumstances. Here are some of his figures: carrier-pigeons, 30–36 miles per hour (over 60 has been recorded, but possibly only with a strong favourable wind); crows, 31–45; small song-birds, 20–37; starlings, 38–49; ducks, 44–59; he also quotes the case of a flock of swifts flying at 6,000 feet above Mosul, in Mesopotamia, which in their ordinary flight easily outpaced the observer's aeroplane when it was doing 68 miles per hour. The air speed of this astonishing flyer is, when accelerated, probably well over 100 miles an hour.

As regards altitude, it seems that although birds have occasionally been recorded as high as 15,000 feet, they are indeed rarely met with above 5,000 feet, while the greater part of flight, including migration, probably takes place within 3,000 feet of the ground.

§ 5

The power of flight has given birds the key to one kind of habitat after another that might otherwise have proved to be too dangerous or too inhospitable. To the conditions of these different haunts, and, in particular, to different modes of procuring food, we see a great wealth of adaptations: there are hunters and fishers, catchers of insects and harvesters of seeds, eaters of crustaceans and eaters of worms, plant-eaters and honey-suckers, scavengers of carrion, and many a "picker up of unconsidered trifles."

The Hunting of the Peregrine

Pride of place may be given to the hunters, and, as a type of them, to the Peregrine Falcon, described by the late Professor Alfred Newton as "the most powerful bird for its bulk that flies." It is a strong fierce bird with long pointed wings, spending no time on its comings and goings and dealing death in mid-air with relentless talons; in spite of game-preserving it still maintains its place as one of the most splendid of native British birds. Its prey consists mainly of other birds, and these it attacks in flight, "stooping" always from above, and killing, not by force of impact, but by the sheer grip of its claws. "Having arrived within a few feet of its prey," wrote Audubon of the almost identical Duck-Hawk of America, "the falcon is seen protruding his powerful legs and talons to their full stretch. His wings are for a moment almost closed; the next instant he grapples the prize, which, if too weighty to be carried off, he forces obliquely to the ground, sometimes a hundred yards from where it was seized, to kill it and devour it on the spot. Should this happen over a large extent of water, the falcon drops his prey and sets off in quest of another. On the contrary, should it not prove too heavy, the exulting bird carries

From "Our Common Sea-Birds" (Country Life, Ltd.).   *Photo: J. C. Douglas.*

HERRING-GULLS (*Larus argentatus*)

The Herring-Gull gets its name from its habit of following shoals of herrings. The Gulls will follow shoals of small fish swimming near the surface. They usually swoop down upon them on a flat curving course, not submerging the body but merely immersing the head and bill.

Photo: Royal Scottish Museum.

GOLDEN EAGLE

The largest of the British birds of prey. Soaring overhead, it is of majestic appearance. The young ones are taught how to hunt and how to kill as well as how to carry and skin their prey.

RAVEN

The brainiest of all our birds.

SECRETARY BIRD

The name is derived from the tuft of feathers looking like a quill stuck behind the ear. The bird lives in Africa and preys on small animals—notably snakes—which it kills with a swift stamp of one of its powerful legs.

it off to a sequestered and secure place." A peregrine can indeed carry a weight almost equal to its own, and a pair nesting on the Bass Rock, in the Firth of Forth, have been known to bring grouse and pheasants from the mainland across two or three miles of sea.

The Peregrine Falcon belongs to the aristocracy of the bird world. It has a haughty stare, a regal dignity, is absolutely fearless, has great reserve power, and, as Mr. Hudson says, possesses a courage commensurate with its strength, and in hunting an infallible judgment. It is one of the most perfect of winged creatures, "so well-balanced in all parts, so admirably adapted for speed, strength, and endurance." The lordly falcon is "the terror of the skies."

"Sooner or later the day always comes in early autumn to birdland when the peewits, feeding in silent battalions together, and the gulls, watching impatiently to rob the peewits of their worms, suddenly arise and wheel in wild disorder to the horizon; when the clustered partridge coveys crouch, like clods to the earth, and the flocks of small birds, feeding in the open fling themselves like a shower of stones into the nearest hedge; when the blackbird issuing from cover turns before he has flown a yard, and darts back again with a chatter of alarm; when, save for the distant cawing of rooks perched on lookout trees, a parish apart, sudden, perfect stillness holds the landscape. Then the peregrine falcon passes, smiting her way from horizon to horizon, and spreading terror as she goes. Who gave the first warning of her coming it is hard to tell. Possibly it was a rook. But the marvel is that the majority of the birds, being young ones of the year, can never have seen a falcon before; yet they fling themselves wildly to right and to left long before the speck in the far skies reveals itself to human eyes as a bird of prey." [1]

The Golden Eagle is the largest of our native birds of prey. The well-known lines of Tennyson spring to the mind:

[1] E. K. Robinson, *The Country Day by Day.*

He clasps the crag with crooked hands;
Close to the sun in lonely lands,
Ring'd with the azure world, he stands.

The wrinkled sea beneath him crawls;
He watches from his mountain walls,
And like a thunderbolt he falls.

The Golden Eagle looks well after its young, feeding them at dawn and dusk each day. "The Grouse that are brought to the eaglet are plucked and headless; the Hares and Rabbits are skinned and made ready in a larder distant from the nest; the youngsters get only digestible food, being unable for some weeks to form pellets." The eaglets are taught how to hunt and how to kill, as well as how to carry and skin their prey. When they are about five months old they are driven away.

The Fishing of the Cormorant

Very different from the habits of these birds of prey is the under-water hunting of the Cormorant, a bird of much less noble habits and aspect, which is notable for clumsiness in the air, and for uncouth appearance on land, as well as for the foul stenches of its untidy nest! Under the water, however, it is a thing of beauty, so perfectly adapted is it to the swift and dexterous pursuit of its active prey. In a tank with glass sides we may see this to great advantage, and note how the wings are kept close to the body—not used for swimming as in the case of penguins and auks—and how the air-bubbles cling to the feathers like bright jewels or polished silver. We can see, too, how the strong hooked beak is used to seize the fish, which is then borne to the surface to be tossed in the air, recaught, and swallowed, for the Cormorant does not swallow under water like a penguin. The Chinese train cormorants to catch fish for the market, a collar round the neck preventing the birds from swallowing their prizes; the same thing was done in Britain at one time, although only for sport.

Different Fishing Methods

It is interesting to compare the different methods of fishing adopted by two of the Cormorant's relatives, the Gannet and the Pelican, and the different forms of beak which go with each. The Gannet, or so-called Solan "Goose," nests in great colonies on several of the rocky islets around the British coasts, and it may also be seen at most times off many parts which are far from these breeding stations. It is a bird of fine white plumage and noble flight, which, soaring at a height and then suddenly dropping like a plummet, uses its long straight beak to transfix fish swimming near the surface.

The Pelican, again, is a fisher of the shallows which wades through the water with its enormous gape at full extent, and the great pouch below its beak ready to receive what comes. A party may work in concert, sweeping the pool in a long line like a living seine net. "The Cormorant pursues, twists, turns, and seizes; the Gannet soars, plunges, and spears; the Pelican sweeps and engulfs."

§ 6

The Wisdom of the Raven

We may refer here to the Raven. Like some of the larger birds of prey, the Raven takes a wife for life, and they use the same nest year after year. As an inland bird the Raven is now not so frequently met with, for it has been driven by persecution from many of its former mountain haunts. Luckily it is one of the hardiest of birds and can adapt itself to great extremes of temperature.

The Raven, the biggest of our Crows, is the brainiest of all our birds. "His family are the great legal fraternity among birds; nimbleness of wit mingled with audacity characterise them all, so that the very first time that I observed the hoodie crow at home I was struck with his laughable resemblance to a barrister in wig and gown. There was the same keen eye for the shortcomings of others, and the general look of mental superiority to ordinary folk." [1]

The Raven has the reputation of being one of the longest-lived birds; it enjoys a reputation also for mimicry. If you climb to its roosting-place on some mountain precipice you may hear "in the silence of the hills how the ravens croon themselves to sleep, uttering reminiscences of the sounds they have been listening to throughout the day." Mr. F. B. Kirkman, in *The British Bird Book,* writes: "From the growing congregation on the ridge there descended through the thickening dusk the strangest of evensongs —a weird, wild medley of many sounds: the barking of dogs, the bleating of goats, the lowing of cows, the becking of grouse calling across the moorland, and now and then the deep-belling challenge of the stag." Their intelligence is almost uncanny, and when we think that they are of savage character and have a deep, harsh, human-like voice, we can imagine some explanation of the evil reputation of the bird, and the sombre superstitions associated with it.

§ 7

Social Life

It has to be confessed that we have a great deal to learn about the inner life of birds. It is difficult to get mentally in touch with them; they have evolved on a different plane from our own. Our sense of kinship with animals is still something novel, but it is ever widening and deepening as we view it more closely and with clearer vision: may we not claim this as one of the steps in the progress of Evolution?

With birds, as with mammals, there are many phases of social life. Some species of birds are more social in their relationship than others; in some there is a more advanced state of community than others. With individuals there may exist mutual friendship; companionship between two birds of the same species, or even between birds of different species, is often seen.

[1] Francis Heatherly, *The Falcon at the Eyrie.*

Photo: "Our Common Sea Birds" (Country Life, Ltd.).

PUFFINS

The Puffin's beak is of singular appearance. After the breeding season its surface peels off in horny plates—shed like the deciduous bark of certain trees—a process parallel to the moulting of the plumage.

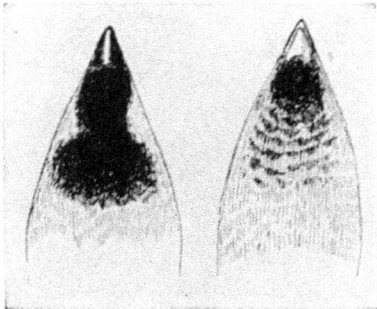

The throat of the House-Sparrow has a very different appearance in summer (left-hand figure) than in winter (right-hand figure). This is due to abrasion; the light tips of the feathers are shed in spring, thus revealing the dark portions.

Photo: James's Press Agency.

MAGPIE

HEN NIGHTINGALE

The mate of our most famous songster.

WHINCHATS

From its perch on the top of the hedges or on low trees, or whilst hovering in the air, may be heard the Whinchat's curious short whistle, *thu-tick-tick*, the latter two notes being exactly like the sound made by hitting two small pebbles together.

The helping instinct is characteristic in birds as in other animals; it is often touchingly human-like. We see it most often in parental care and in the feeding of each other by the sexes, but it is shown frequently in other ways. Mr. W. H. Hudson, speaking of the Military Starling of the pampas—a bird of social disposition—tells this story: "One day I was sitting on my horse watching a flock feeding and travelling in their leisurely manner, when I noticed a little distance behind the others a bird sitting motionless on the ground and two others keeping close to it, one on each side. These two had finished examining the ground and prodding at the roots of the grass at the spot, and were now anxious to go forward and rejoin the company, but were held back by the other one. On my going to them they all flew up and on, and I then saw that the one that had hung back had a broken leg. Perhaps it had not been long broken, and he had not yet accommodated himself to the changed conditions in which he had to get about on the ground and find his food. I followed and found that, again and again, after the entire scarlet-breasted army had moved on, the lame bird remained behind, his two impatient but faithful companions still keeping with him. They would not fly until he flew, and when on the wing still kept their places at his side, and on overtaking the flock all three would drop down together." As Mr. Hudson says, it is possible to mistake for friendship an action which, at all events in its origin, is of a different nature.

Instances of such altruistic behaviour, to be attributed to the *helping* instinct in animals of social habits, are common. Mr. Frank Finn relates that the upper bill of a Huia, an insectivorous bird of New Zealand, by some accident or natural deformity had grown into the shape of a corkscrew, and it was not apparent how it could get enough food to support life naturally. It seems it had been fed for some time by a devoted mate.

The Social Habit

The development of a social habit at the breeding season is a well-marked characteristic of many kinds of birds, and it is by no means confined to those which are gregarious at other times; conversely, it is also true that some birds which at other seasons band together are among the least social at this special time. More than one factor is probably involved: the scarcity of suitable sites—for marsh-fowl, for example—may be a reason for concentration in special spots, and strength of defence against enemies may often be an advantage gained. In other cases the problem of food-supply will tend to produce distribution rather than concentration, and this is especially the case with many of the smaller species of our common birds: among warblers, for example, there is a marked tendency for a pair to select a small territory within which they will remain and from which they will endeavour to exclude all other members of their own species and even, in due course, their own young.

Many birds, like human beings, would seem to enjoy the company of their kind. The gregarious habit is common, for example, among rooks, starlings, pigeons, swallows; parrots roam in bands, apparently for the pleasure of one another's company.

We may have crowds and associations, however, without sociability; a community of separate individuals may exist without there being any corporate life or power of acting as a unity. Still, we do see many instances of a capacity for unified action and distinct features of a social life. "There appears to be an intellectual advantage in sociability, if we may argue from the fact that many social animals show a high development of wits. The three cleverest kinds of birds are rooks, cranes, and parrots, and they are notably social. There is, of course, the danger of putting the cart before the horse, for it may be that the sociability is in part the expression of good brains. It may also be argued that the non-gregarious crow is just as clever as the social rook, and many analogous instances might be given." [1]

The Rook is the best example of our gregarious birds. There is no doubt that the members of the Crow family have fine brains,

[1] Professor J. Arthur Thomson, *The Wonder of Life.*

and great power of vocalisation which may develop to a remarkable extent. Experts tell us that the Rook has command of between thirty and forty notes. To learn to what extent they employ them one has only to listen to "the black republic in the elms," after the breeding season is over.

Professor J. Arthur Thomson, in *Secrets of Animal Life,* says: "Like many creatures well endowed with brains, rooks exhibit what must be called play. There are gambols and sham fights, frolics and wild chases, in which, curiously enough, jackdaws and lapwings sometimes become keenly interested. But who knows the real truth about rooks posting sentinels, which is so often alleged? Who knows the significance of the vast congregations that are sometimes seen, and who can tell us if there is any truth at all in the alleged 'trials' of individuals who have defied the conventions of the community? . . . But the central interest is in the rooks reaching forward to a communal life with certain conventions, and to the crowded nest in which we see the beginning of a continuous social heritage of objectively enregistered traditions." There may be far over a thousand nests in a rookery, and the same site may be used for more than a century.

Rooks certainly have a considerable vocabulary. There is not, indeed, any *language* in the strict sense—man has a monopoly of that; but the rooks have *words* just as dogs have, definite uttered sounds which have definite meanings. We hear the rooks use certain words when we move suddenly beneath the trees, and other words are uttered when a bird intrudes on its neighbour; there is a word when the rook sinks down upon the nest, and another word when it flies clear of the rookery and makes for the fields. What danger-signals, what scoldings, what chucklings, what exultation, what reproaches, what encouragement do we not hear?

Mutual Protection

Mr. W. P. Pycraft, in his *History of Birds,* says: "Among gregarious species some display a much more intimate association than others—are more social in their relationships. And this is shown very clearly in the devices which some species have adopted for their mutual protection during sleep. The common partridge, as is well known, lives in small companies, or 'coveys,' which scatter only while feeding, and then not far enough to be beyond call. Later in the day, as soon 'as the beetles begin to buzz,' says Professor Newton, the whole move away together to some spot where they jug, as it is called—that is, squat and nestle close together for the night; and from the appearance of the mutings, or droppings, which are generally deposited in a circle of only a few inches in diameter, it would appear that the birds arrange themselves also in a circle, of which their tails form the centre, all the heads being outward—a disposition which instinct has suggested as the best for observing the approach of any of their numerous enemies, whatever may be the direction, and thus increase their security by enabling them to avoid a surprise. Ducks similarly take special precautions to secure safety during sleep, when this must be taken in exposed situations, as when, for example, they desire to doze between the intervals of feeding during the night, which they pass afloat. At such times they keep close together, and to avoid drifting ashore keep one leg slowly paddling, and thus drive themselves round in circles."

There is sometimes co-operation in hunting as we have already noted in the case of pelicans, which combine in a crescent and, wading shorewards, drive the fish before them; when they have got them cornered they fill their huge throat pouches. It is said that a pair of golden eagles will occasionally hunt in concert, one beating the bushes while the other flies overhead, waiting to pounce.

With birds, as with other animals, we see, as we do in human beings, that some individuals are gifted above others of their kind. A few may have a keener sense, greater strength or power of leadership, a more helpful spirit than their fellows. This counts for much in a social state. "The action of the gander and of the trumpeter in driving their fellows home in the evening must be regarded as similar in its origin to that of the male swift, when he hunts his mate back to the nest, and of the sand-martin I observed chasing the females of the colony to their burrows. In a lesser way it may be seen in any flock of birds; they move about in such an orderly manner, springing, as it appears to us, simultaneously into the air, going in a certain direction, settling here or there to feed, presently going to another distant feeding-ground or alighting to rest or sing on trees and bushes, as to produce the idea of a single mind. But the flock is not a machine; the minds are many; one bird gives the signal—the one who is a little better in his keener senses and quicker intelligence than his companions; his slightest sound, his least movement is heard and seen and understood and is instantly and simultaneously acted upon."

Interrelations

Many curious associations are formed by birds during the breeding season. The Puffin is quite capable of making a hole for itself in the face of some precipitous slope, but frequently it prefers to appropriate a rabbit's burrow, ejecting the rightful owner without ceremony. Other burrowing birds are often more accommodating, for the Burrowing Owls of America live amicably with the Prairie Dogs whose retreats they so often share, and in New Zealand the same holes are shared by Petrels and Tuatera Lizards without apparent friction. In cases of this kind, however, it is always possible that the partnership has other advantages— such as common defence or watchfulness—than the mere saving of labour on the one hand or on the other: there is the curious case, for instance, of the Ruddy Kingfisher of Borneo, which makes its nest in the hive of a peculiarly vicious kind of bee!

§ 8

Our Common Birds

The late Professor Newton has an interesting passage in which he shows that we can tell which birds were most familiar to our forefathers by their having a pet name added. Thus the Daw is the Jack Daw, the Redbreast is Robin, the Wren is Jenny, the Pie is Magpie, the "Mag" being short for Margaret. In early prints of ploughing, the closeness of the connection between men and birds is naively pictured. In one of the earliest illustrations of sowing, for instance, the birds crowd so closely on the heel of the sower that they have to be driven off with stones or even whips, and they are seen springing beyond the leap of a small dog that has been sent to chase them. In modern times the charm possessed by birds is partly that of friendship, but more that of delight in their songs and feathers. The following birds form only a few examples chosen for some special interest.

The Nightingale

It has often been disputed whether the Nightingale's song is really the sweetest. It certainly owes something to the stage on which it is set, for when the bird arrives the field and garden are gay with spring flowers. The Cuckoo arrives just about the same time. It sings all day, but the Nightingale mostly in the evening, and the sweetness of his note is enhanced by the light of stars and the scent of blossom. Whether it is a melancholy or a merry song has long been disputed. It certainly is not loud, because when the Nightingale sings by day it is not noticed amid the clamour of other bird music. Mr. W. H. Hudson says: "Its phrasing is more perfect than that of any other British melodist, and the voice has a combined strength, purity and brilliance, probably without a parallel."

The Blackbird and the Song-Thrush

The Blackbird's voice is remarkable for its great strength and for the wonderful rich quality of its tone. He is a clever mimic, like several other songsters, and has been heard to imitate the Nightingale's song with some measure of success. There are several recorded instances, too, of his crowing exactly like a

Photo: Oswald T. Wilkinson.

JACKDAWS

One of the most entertaining and amusing of all British birds. Its passion for mischief is unexampled in the feathered world.

MEADOW PIPIT CLEANING ITS NEST

It picks up the excrement of the young, which is enclosed within a gelatinous capsule, and carries it off to a distance.

HERON (WITH YOUNG)

The Heron, rising in the air and fleeing from its enemies, the Falcons, "will, in its efforts to lighten itself, and so keep *above* its pursuers, disgorge any food which may still be undigested. So that in watching such an flight one might expect to see a small shower of—for instance—fish and shrimps fall from the clouds."

domestic cock—"apparently enjoying the sound of the responses made by the fowls of the neighbouring farmyard"—and of his cackling like an egg-proud hen! Some prefer the song of the Blackbird to that of the Thrush. It certainly is the sweeter of the two, but it is not so long continued. It may vary with the district, and some hold that the Surrey Blackbird is the sweetest songster of his kind. The period of song is identical with the visit of the most delicately beautiful of all butterflies—the Orange-Tipped. Even the little Song-Thrush, a close relative of the Blackbird, is a louder and more persistent singer than the latter, although in that respect he does not compete with the larger Missel-Thrush, which can often be heard pouring out his bold loud notes from the topmost twig of a bare tree in the month of January. The song is in keeping with his character. Mr. W. H. Hudson thinks that "The Throstle is by far the finest songster. His chief merit is his infinite variety. His louder notes may be heard half a mile away on a still summer morning, his lowest sounds are scarcely audible at a distance of twenty yards. His purest sounds, which are very pure and bright, when contrasted with various squealing and squeaking noises, seem not to come from the same bird. . . . As a rule, when he has produced a beautiful note he will repeat it twice or thrice." While the Blackbird is cunning and secret in his ways, creeping round the roots of the yews and other shrubs, the Thrush boldly roams across the fields.

The Lark

The songster most closely associated with the farmlands is undoubtedly the Lark. He is the earliest rising of all the birds, and when in full voice, as he is just about the time when the young wheat is tall enough to cover him, he may be heard pouring out his song before sunrise. He is not one to confine his charms to his courting days, but has been heard in every month of the year except September, his moulting time. It is in Spring

and early Summer, however, that he pours forth his best music.
The song has words for it in the folklore of many countries, and
the following rhyme succeeds in conveying an idea of it:

> Tu whit, tu whit, tu whee,
> No shoemaker can make boots for me
> Why so? why so? why so?
> Because my heel's as long as my toe, my toe.

The Wood-Pigeon

No voice is more closely associated with the beautiful wooded
landscapes of England than the love-song of the Wood-Pigeon.
According to an ancient legend, the words it tries to say are:
"Tak two coos, Paddy," the legend being that in the Golden
Age the Wood-Pigeon laid its eggs on the grass, but they were
trampled upon by two cows. An Irishman led one away, and
the Wood-Pigeon prays in vain for him to take the other, to
which the Partridge is supposed to reply: "De'il take it"—a wonderfully close imitation of its apology for a song. The Little
Dove, the Turtle Dove, or the Croodling Dove has a sweet short
song that fits in well with the whisper of the summer leaves. It
is an old country saying that when you first hear the croodling
of the Little Doe, then is the time to sow your swedes.

The Bullfinch and the Goldfinch

One has often wondered if there is a manner of accounting
for the different marital qualities that characterise birds. Take
the cock Partridge, and you find a model father—one that will
stand up to anything in defence of his young—while the cock
Pheasant is a very gay Lothario. The most faithful of our birds
is the Bullfinch. The male and female do not only stick together
during the breeding season, as is the case with most birds, but
along the lanes in winter you may see the male and female picking up morsels of food on the black hedgerows. They do not
keep close together, but never go out of hearing of one another,

and it is very easy to imagine words, for the conversation which they keep up. The Goldfinch is perhaps the most beautiful of all the feathered folk in the English landscape. In Autumn it is a very pretty sight to see a little cluster of them feeding on thistledown, and performing the most delicate acrobatic feats in balancing themselves so as to pick it from the plant.

A Few Woodlanders

Variety of character in birds is nowhere more marked than among the more familiar inhabitants of the woodland. Take the Jay—clean-made, bright-coloured, with a voice that is raucous but seems always in tune with the noise which the wind makes blowing through the tall trees. He is a gentleman in appearance, but his flight is as awkward as the gait of a yokel. Moreover, Nature has endowed him with a thieving and lawless character. He steals the eggs from the nest, and makes a meal of any fledgings that he can lay hold of. Yet he is very cunning about concealing himself during the breeding season, when he has to think of the safety of the family as well as his own. For the time being, the loud cry is stilled, and the bird, on being disturbed, shifts slyly and quietly from one tree to another. He has a natural genius for concealing his nest, and in that way differs very much from his relative, the Magpie, whose idea of architecture is simply to pile woody twigs upon woody twigs, so as to make a conspicuous and monstrous habitation. The Magpie used to be a favourite domestic pet, but its numbers have now been greatly reduced, so that to see several of them together, which used to be considered very unlucky, is almost impossible in some districts. They very often go in threes, for some reason which we cannot explain. The Magpie can be taught to articulate a few words; he is inquisitive and loquacious. "The usual sound emitted by the magpie is an excited chatter—a note with a hard percussive sound rapidly repeated half-a-dozen times. It may be compared to the sound of a wooden rattle or to the bleating of a goat, but there is always a certain resemblance to the human voice in it, especially when the birds are unalarmed, and converse with each other in subdued tones." The Heron is a bird of the woodland, in so far as it is there he makes his heronry. It will frequently be found closely adjacent to a rookery, but the two colonies do not always live at peace, although in a case the writer knows of, quite near London, they have done so for many decades. The Rooks are numerous and aggressive, and though an individual Rook could not hold its own with a Heron, numbers usually prevail when a battle royal takes place. In habit the Heron is a bird of the brook and river, and there can be little doubt about his favourite diet being of fish. He loves to stand in a clear, shallow stream, apparently motionless, but should an eel creep out, or a bolder trout try to make a passage up-stream, the Heron's keen eye sees it at once and down comes his beak like a sharp spear, the chances being that the next experience of the fish is that of being borne through the air, to be eventually swallowed and either wholly or partly digested. In the latter case, the process is stopped in order that the young may receive the food in a softened condition.

The Green Woodpecker is a common British species, whose bright plumage is less conspicuous among the trees than might be thought, but whose presence is often betrayed by the loud cry—like a burst of demoniac laughter—or by the strong "tap, tap, tap" of its beak as it sounds the tree-trunks for rotten portions where insects may be found. The woodpecker's strong beak, adapted to its mode of feeding, is well suited also for the work of excavating a nesting hole, and a deep cavity with a small horizontal opening at the top is hollowed out.

The Waterhen and the Coot

The Waterhen looks black at a distance, but on closer observation discloses many charming shades of colour. It is a bird that seems to thrive and increase in numbers more than its com-

Photo: Royal Scottish Museum.

COMMON GULL (*Larus canus*) WITH NEST, EGG, AND YOUNG

Colonies of them are often found on lochs at a distance from the sea. The birds roam over moors and marshland, and they are often to be seen closely following the plough, picking up worms and grubs. Like the Herring-Gull, it has been observed dancing on the sand or mud of shallow pools to force up marine worms from below.

LAPWING OR PEEWIT SETTLING DOWN ON ITS EGGS

If an intruder comes near, the young will take warning from their excited parents wheeling overhead calling "pee-a-weet, pee-a-weet" and squat flat on the ground in absolute stillness. No noise will make them move a muscle so long as their parents call overhead.

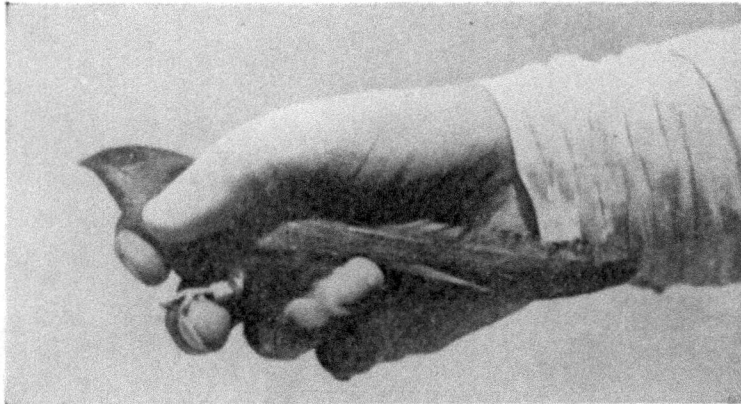

From Smithsonian Report, 1913.

Chimney Swift ringed at Meriden, New Hampshire, U. S. A., in June, which returned to his old chimney after wintering in the Tropics in the June of the following year. Similar records have been obtained in this country in the case of the Swallow, the Swift, the Spotted Flycatcher, and other summer visitors.

Photo: Royal Scottish Museum.

BLACK-THROATED DIVER (*Colymbus arcticus*) WITH NEST AND EGGS

A beautiful bird which nests in some remote parts of Scotland, but is commoner as a winter visitor to our coasts. The nest is usually close to the water on an island in some small loch. The bird's legs are set very far back for purposes of swimming and diving; on land, therefore, it does not stand upright, but pushes itself forward on its breast.

panion, the Coot. Yet it nests often in a perilous position. You may seek for the nest either among the rushes and flags at the border of a stream or on the long willow branches that stretch out close to the surface of the water, if they are not touching it. Country folk believe that in every normal year there is a May flood, and when that comes the water very frequently lifts the nest of the Waterhen out of its mooring and carries it downstream. The faithful bird will go a long distance in its curious little ship, but is compelled to vacate it at last, as such floods carry down the branches of trees, trunks that have been lying on the bank, and a great deal of miscellaneous debris capable of wrecking the poor craft. Not that the Waterhen is likely to suffer personal injury, as she will dive into the strongest running stream and escape scatheless.

The Grebes

The Little Grebe is to be met with on inland waters all the year round. In winter it resorts to rivers and larger bodies of water, when the small ponds beside which it often nests are apt to be frozen over. Its supreme accomplishment is that of diving and hiding itself among the stems of waterplants or other cover. It must, of course, come up, but it is amusing to notice the length of time it will remain under the water, and the distance it will often travel before it makes a second appearance. The Great Crested Grebe is one of the stateliest and most beautiful of our inland water birds.

Visitors from the Sea

One of the most beautiful sights to be seen in this country is that of a colony of Blackheaded Gulls nesting beside a lake or in swampy places far away from the sea-coast and estuaries where they may be found in winter searching for small fishes or other food cast up by the tide. In days of old their eggs were prized as food, and even the young were taken, but the modern palate does not set so much value on them. The movements inland are made with great regularity, the birds appearing at one gull-pond, of which we know, about March 27, scarcely ever a day before or a day later. They raise their young while the Corncrake is singing its mournful and monotonous ditty in the new grass and the growing wheat. A hill country attracts them because of the little streamlets which provide plenty of food. They know as well as the angler does that the trout lie with their heads up-stream, waiting for any little titbit in the shape of a worm or fly which the water brings down. When the Gulls are fishing, one can watch them beating their way up past a succession of gravelly shells into which they occasionally dip for prey. When they come to the end of the beat, they fly back round the shoulder of the hill out of sight of the stream and resume operations where they started before.

Birds of the Moorland

There is no prettier adjunct to a moorland, or a bare field, than a flock of Lapwings. They fly together and alight together in Autumn and Winter when not breeding, but in nesting-time they go in pairs, though usually there are dozens and sometimes hundreds in the same field. The bird is a simple creature in so far that its nest is little more than a slight hollow on the bare earth. In Spring they can be seen sitting on their eggs without making the slightest attempt at concealment, so that the individual who goes out to collect their eggs need only march up to a sitting bird, but if it rises he must keep his eye on the place from which it springs. There never can be much doubt as to whether or not the nest is close, because, if it is, the birds shriek and swoop at the intruder, as if they were going for his head or eyes. Should an animal other than a man come, they will indeed carry out the threat. No sooner are the young out of their shells than they begin to run, and if chased, will select a hiding-place. It may be close by stones as grey as themselves, or in the short herbage which early Spring brings with it. A trained eye is needed to distinguish them from their surroundings, even at a short distance.

The Curlew haunts the sea-shore during the greater part of the year, but in Spring retires to some slack or valley in hilly country, and makes a nest on the ground. The situation is generally very lonely, and the watchful birds quickly show themselves alive to the presence of a stranger. Usually, their note is a monotonous and melancholy sound, heard, as it often is, at night-time in the stillness of the moorland, but we know of no other bird that makes the clamour the Curlew does when its domestic privacy is invaded. It flies up and down the valley, shrieking to awaken the echoes, and looking as if it would like to do something dreadful to the human who had ventured into its domain.

The Snipe is the most difficult of indigenous game-birds to shoot, on account of its trick of half-stopping and suddenly darting. During the breeding season he performs curious antics in the air, rising to a great height, and "precipitating himself downwards with astonishing violence, producing in his descent the peculiar sound variously described as drumming, bleating, scythe-whetting, and neighing." The peculiar drumming sound was long the subject of controversy, but recent observations have made it clear that it is due to the vibration of the two outer tail-feathers, which have a peculiar structure.

§ 9

The Cuckoo

The Cuckoo, as is well known, not only builds no nest of its own, but foists its eggs on other species, and has its young reared without trouble to itself but to the great detriment of the rightful children of the foster-parents. The story, indeed, is one of the most curious in the whole realm of natural history, and the facts are now becoming better known; among other new evidence, the recent intensive observations and wonderful cinematograph records of Mr. Edgar Chance have placed several points beyond doubt

The Cuckoo's Procedure

It seems to be the case that each female cuckoo has its chosen territory of operations and that deliberate choice of nests is made in advance of the date of laying. When the time for laying comes, the selected nest is approached, the cuckoo takes an egg from the nest in its beak, settles on the nest, lays its own egg, and then flies away with the stolen egg, which it either eats or drops at a distance. The whole manœuvre takes but a few seconds and may be carried out despite the frantic efforts of the small and unwilling hosts to drive off the intruder. Sometimes the procedure varies, for no cuckoo could lay in a wren's nest, for instance, and in cases of that kind the egg must be laid outside and inserted with the beak. The point of principle, however, is that the cuckoo certainly does not fly about carrying an already laid egg and looking for a suitable nest to victimise.

The Young Cuckoo's Part

One cuckoo does not normally lay two eggs in the same nest, but different cuckoos may chance to select the same victim if there has been an encroachment of territory. Once the act has been accomplished the foster-parents do the rest until the eggs hatch out; then begins the second part of the Cuckoo's villainy, for the young foundling has in his earliest and comparatively helpless days the inborn habit of removing the other chicks from the nest by getting his back under them and heaving them overboard. So it happens that the foster-parents are soon left with but one charge, whose voracity keeps them perpetually busy and whose body speedily fills up the nest. Still the poor dupes go on feeding the parasite, even when he is much bigger than they are; one of Mr. Chance's photographs shows a bloated

Reproduced by courtesy of Messrs. Methuen & Co. from "A History of Birds," by W. P. Pycraft. (After a painting by G. E. Lodge.)

THE AMHERST PHEASANT IN DISPLAY

A fine example of ornamental plumage, lending itself to display in courtship. The male has this garb all the year round, but the hen is of sober hue. Like several other gorgeous species such as the Golden Pheasant and Swinhoe's Pheasant it is a native of China.

young cuckoo sitting on a post, while the much smaller pipit dutifully feeding him must needs stand on his shoulder, so to speak, for the purpose! The whole story is one of effective adaptation on the part of the Cuckoo and of the weakness of blind instinct on the part of the foster-parent.

The most interesting theoretical point about the Cuckoo has to do with the colour of the eggs, which is very variable, but tends to be like that of the eggs of the chosen foster-mother. That one hen cuckoo always lays the same type of egg seems to be thoroughly established, but it is still a matter of speculation whether the character is hereditary and, if so, in what manner.

The Cuckoo victimises a large number of different species as foster-parents for its young, but all the usual ones are small insectivorous birds. The degree to which the Cuckoo's egg resembles the others varies greatly; sometimes there is almost a perfect match, at least in colour, but in other cases the similarity is slight or even non-existent.

§ 10

MIGRATION

The scientfiic investigation of migration is greatly complicated by the difficulties of making observations. It is not now believed that the greater part of migration takes place at immense altitudes, and at an accelerated rate of flight which makes enormous journeys possible for birds in a single night. Nevertheless it remains true that a great deal of migration is nocturnal, and that, for other reasons also, it is difficult to observe. At certain times and places, however, much migratory flight can be actually observed. We have, for example, this recent description of the passage of swallows on Heligoland: "All through the forenoon, as we sat in the autumn sunshine near the narrow northern apex of the island, the swallows came in over the sea from the northeast in the teeth of a southerly gale. No large flocks were seen;

but, on the other hand, scarcely a minute elapsed without the arrival of a fresh party of from half-a-dozen to a score of birds. They seemed to fly low over the sea, but rose as they approached to the level of the cliff-tops. We could not make them out at any distance, for the observer can find no worse background for small birds than grey, moving water. The stream was continuous and the direction unvarying, so far as we could judge. Each party rose to our level on the top of the north point, flew unhesitatingly along the western side of the island, and disappeared again at the south-western corner. Not one in a hundred quitted this line or stopped to circle round; none seemed inclined to break their journey so early in the day, in spite of the contrary elements. The whole was for us just a momentary peep at one of the countless tiny channels by which the bird-life of northern Europe was then ebbing southward."

Migration of Starlings

A more comprehensive idea of migration is obtainable from Gätke's observations extending over the whole length of a season; let us summarise the diurnal movements of starlings, as observed by him on Heligoland during the autumn of 1878. Early in June came a few old birds in worn plumage, birds which had probably remained unmated or had early lost their broods. On the 20th June came the first great flights of young birds of the year, migrating by themselves in advance of their parents although only a few weeks out of the egg. These youngsters continued to pass till the end of the month to the extent of thousands daily. In early July the daily passage was estimated in tens of thousands, and on the 25th the movement closed with the passage of "immense multitudes." Two months followed during which no starlings, young or old, were to be seen. On the 22nd September old birds, now in fresh plumage, passed in flights of many hundreds. During October the flights increased to thousands, and on the 14th the movement reached a climax with "starlings in hundreds of

Photo: F. R. Hinkins & Son.

Portion of tip of downy feather from the under tail-coverts of the Missel-Thrush. (Highly magnified.)

Photo: James's Press Agency.

SWINHOE'S PHEASANT (MALE)

Another example of brilliant plumage lending itself to "display" during courtship.

BLACK-COCK ON THEIR PLAYING-GROUND

Photo: R. Chislett.

A PAIR OF GREAT SKUAS AND THEIR YOUNG

The Great Skua is a species related to the Gulls on which it preys by compelling them to disgorge the food they have just obtained. The similar Arctic Skua, to lead unwelcome visitors away from the neighbourhood of its nest, will cunningly sham a broken wing or leg, luring the intruder to give chase.   When at a safe distance from the nest the bird will suddenly rise in the air and fly away.

thousands." By the end of the month the great flights had ceased, but through November, and even up to the 18th December, the birds continued to pass in "daily flights of from forty to sixty individuals."

Lighthouse Scenes

Nocturnal migration is also often observed at lighthouses and lightships, and especially when the weather is foggy thousands of birds, dazzled by the lantern rays, dash themselves against the glass. As Gätke says: "The whole sky is now filled with a babel of hundreds and thousands of voices, and as we approach the lighthouse, there presents itself to the eye a scene which more than confirms the experience of the ear. Under the intense glare of the light, swarms of larks, starlings, and thrushes career around in ever-varying density, like showers of brilliant sparks or huge snowflakes driven onwards by the gale, and continuously replaced as they disappear by freshly arrived multitudes. Mingled with these birds are large numbers of golden plovers, lapwings, curlews, and sandpipers. Now and again, too, a woodcock is seen; or an owl with slow-beating wings, emerges from the darkness into the circle of light, but again speedily vanishes, accompanied by the plaintive cry of an unhappy thrush that has become its prey."

The modern method of marking with numbered aluminium foot-rings has already added greatly to our knowledge of the actual journeys performed by individual birds. By this means, for instance, white storks marked in the nest in East Prussia have been traced south-eastwards across Europe to Syria, Palestine, and Egypt, and thence up the Nile to Lake Victoria Nyanza—and also away eastwards near Lake Chad, in the very heart of Africa—and so southwards through Rhodesia to Natal, the Transvaal and Cape Colony. Five separate swallows, marked with aluminium rings in this country, have been found in South Africa in winter.

Migrations of Lapwings

Many birds, such as lapwings, or peewits, marked in Scotland as chicks in summer have been recovered in winter from Ireland. Other Scottish lapwings have wandered further and have been recorded from the west coast of France or from Portugal. In a few cases, too, lapwings were reported during the winter months from their native districts. It is therefore evident that even within a single species, in a single area, there may be resident and migratory individuals, and, among the migrants, some which go much further afield than others.

There is no doubt that many birds on their return make for their birthplace. A swallow marked when it was a young one has been found thus to return to its native farmyard. Birds would appear to return in spring impelled by a greater urgency than in the autumn migration, when we see sometimes a good deal of dallying. Some birds are known to make trial trips and begin their journey with short stages. On the return, some authorities believe there is evidence that the spring journey is more direct, that short cuts are found, and that haste is evident. When weather conditions are very bad there is often great loss of life. "The streets of towns are sometimes strewn with thousands of birds that have gone astray and have perished in the cold. As many as five hundred nightingales have been gathered in a single day from one small town."

§ 11

The Purpose Served by Migration

Migration must serve some good purpose and be of advantage to the species which possess the habit. It is, indeed, an expensive habit, involving the perpetuation of a complex instinct, the output of a large amount of energy, and the facing of great risks and a heavy mortality: these factors would surely be enough to wipe out a species in the keen struggle for existence did not some great compensating advantage also accrue. For the departure of many birds on the approach of winter we can perhaps see good reason, probably not so much in mere cold itself, but in the decrease in food-supply, in the freezing of ground and water, and in the shortened hours of daylight. In the return from the south in the spring we may see an expression of a need for expansion during the breeding season—to obtain more room, abundance of nesting-sites, and fresh sources of food-supply.

We must distinguish carefully between reasons and causes of migration—between "why?" and "how?" Valid although the reasons given may be, they do not in the least explain *how* the migratory habit has come to be; to miss this point is to fall into the trap of imagining birds as endowed with human knowledge and intellect—with the power of adopting a reasoned course of conduct, based on the foreknowledge of seasonal events and on an appreciation of geographical differences.

Causes of Migration

Two points must strike us as being significant. One is that migration is a very regular phenomenon, happening year after year according to the same pattern, without marked differences corresponding to annual variations in climate, and showing none of the features to be expected in an emergency effort created anew each season. Secondly, much migration takes place long before it seems to be necessary, for in the British Isles southward movements begin as early as July. Many migrants, too, go further than seems to be required, overshooting the mild winter of the northern subtropics to find a similar climate in the summer of temperate regions of the Southern Hemisphere.

The conclusion seems inevitable that migration is a very old habit, an inborn instinct which was developed ages ago, and which manifests itself year after year in a uniform manner and without any remarkably close conformity to immediate conditions. For an explanation of this ancient origin of the instinct we should doubtless look to the former history of birds for some more compelling circumstance capable of initiating the habit which is still maintained. Some have supposed that the last Glacial Epoch, or Great Ice Age, may have driven birds gradually southwards, and after a long time allowed them to return gradually northwards: but during the second phase, it is thought, they would come north only for the summer and return in between to the alternative home they had learnt to know. Others have imagined birds as originating in the south and gradually extending their range in search of fresh feeding-grounds for the hungry mouths of the breeding season, going further and further each summer, but always returning in winter to the original cradle of the race.

If we admit that the immediate seasonal changes are insufficient in themselves to cause migration, beginning so early in each autumn as it does, we must yet invoke them to some extent to complete the other theory. If migration is an ancient habit, annually reborn, there must still be some immediate factor stimulating the latent instinct. Events not in themselves of sufficient strength as causes may yet serve to release more powerful energies, just as a detonator explodes the bursting-charge: so many subtle changes, either in the seasons or perhaps in the functional cycle of the bird's life, awaken the compelling instinct which causes birds to cross unknown seas and continents in accordance with some ancient plan.

How do Migrants Find their Way?

What routes do migrants follow, and how do the birds find their way? We must remember here, again, that migration is in the main a very orderly phenomenon which takes place year after year according to the same pattern. We have now evidence, too, that as regards summer quarters, at least, it is common for birds to return to the same places with great accuracy. Any suggestion, therefore, of a mere haphazard movement with a vague general direction may be dismissed as being inconsistent with the facts as we know them. Other points to be remembered are that much migration takes place at night, and that wide stretches of open sea are habitually crossed. Furthermore, the young of the year in many species migrate southwards before the parents—in the case of the Cuckoo, long after their parents—and must thus find their way without any memories to guide them. Anything which lies in the experience of the race, as distinct from that of the individual, must in these cases be handed on by inheritance purely and not by tuition and imitation.

Our knowledge of the routes that birds follow in their migratory flights is still very scanty. Hooded Crows caught and marked as birds of passage at the south-eastern corner of the Baltic have been shown to come from Southern Finland and the Petrograd district of Russia, and to follow the coasts southwards and westwards as far as the north-eastern corner of France. Black-headed Gulls ringed at the same place, but as nestlings, have been reported from right round the coasts to the Bay of Biscay, from along the courses of the Rhine and the Rhone, and as far as the Balearic Isles, and from along the courses of the Vistula and the Danube and across to Northern Africa.

In its migratory flight the whole life of a bird is raised to a higher pitch. It is estimated that many birds attain a speed of fifty miles an hour, and a carrier-pigeon has been known to keep up the rate of fifty-five miles an hour for four successive hours. It is unlikely that this is often surpassed by migratory birds on long-distance flights.

"Homing"

The question "How do birds find their way?" is not one which can be answered at present. More must first be learnt of the nature of the routes which are in fact followed by migrants, of the relationship of particular summer quarters to particular winter quarters, and as to whether winter quarters are as clearly defined and as accurately sought out as summer quarters are known to be. It is probable, however, that the question may be narrowed down by the elucidation of that special acuity of the senses, or whatever it may be, which underlies the "homing" capacity so well known in birds. Recent experiments by Professor J. B. Watson and Dr. K. S. Lashley have had as their subjects the Noddy and Sooty Terns nesting on the Tortugas Islands in the Gulf of Mexico. Birds taken from their nests and transported by ship in closed cages were shown to be capable of finding their way back from Galveston (to the east) or from Cape Hatteras (to the north), distances of over 850 miles, or from intermediate points at sea entirely out of sight of landmarks of any kind. In being taken northwards, too, the birds were removed beyond the limits of the species' natural range, and the absence of any previous experience in that direction was all the more certain. At least, therefore, we must concede a very highly developed "sense of direction" or "bump of locality."

§ 12

PLUMAGE, COURTSHIP, AND MATING

It does not come within the scope of this work to go into the question of the general classification of birds, neither can we consider in detail the characters of bird structure or of feathers and plumage. A bibliography is given at the end of this chapter which will be useful for readers who wish to have more information on these interesting subjects. A volume might be written on any one of them. We cannot pass over altogether, however, the nature of feathers and plumage.

The acquisition of feathers must have been one of the great steps in the progress of birds towards their present position as the supreme flying animals *par excellence*. It is indeed but to forge another link in that evolutionary history to find that feathers are modified scales and therefore closely akin to the typical covering of reptiles. Let us notice, too, that the unfeathered parts of a bird bear ordinary scales, the one form, as it were, simply replacing the other where it is more suitable. The scales on the toes are often suggestively reptilian in appearance, and when there are also feathers about the toes they grow not *on* the scales but from *between* the scales—from between the other scales we may indeed say to emphasise the point.

Plumage Coloration

The feathers of many birds are richly coloured, and even those of sober hue may be very beautifully marked. In some cases the colours may be due to actual pigment; but in others, especially blues and greens, the minute physical structure of the feathers is responsible and wonderful effects of iridescence are produced.

Brilliance of plumage is often associated with the mating season, but this is far from being a general rule. In some instances the male has a special breeding plumage, and sometimes both sexes have this, examples of each kind being found among the Plovers. In other cases the male has brilliant plumage for most of the year, like the Mallard, while his mate is always dull. In many species, on the other hand, the sexes are alike and have a similar appearance all the year round; this permanent plumage may be dull-coloured as in the Song-Thrush or Curlew—wonderfully beautiful birds, nevertheless—or brilliant as in the Kingfisher. Most birds that have a permanent bright plumage, however, are dull in their first year, as is the case with the afterwards splendidly iridescent Starling, but in some cases, such as the Kingfishers and the Parrots, the gorgeous plumes have appeared before the birds leave the nest. One other kind of change must also be mentioned, namely, the seasonal changes of the Ptarmigan, which is white during the season of snow and of duller appearance when its native hills are brown once more.

Courtship and Mating

Some of the most interesting habits of birds are those associated with the mating season. In many cases there are curious ceremonies of courtship, often with wonderful "display" of brilliant plumage or with great exuberance of song, and sometimes there are fierce fights between rival males. The Peacock spreads and erects his magnificent train, the Argus Pheasant displays long plumes on his wings as well as on his tail, and the different Birds of Paradise glow with gorgeousness in their almost every feather. Many a relatively "dowdy" bird—as judged by human eyes—may also be seen posturing in much the same way as his more ornamental brethren, and we must be chary of denying to any bird strange beauty in the sight of his love!

In the ordinary Black Grouse we may find a habit of display as well marked as that of any inhabitant of tropical jungles; it gives, indeed, an example not only of individual display but also of a collective "tournament" in which rival blackcocks strive to impress the greyhens which they wish to win as mates. In Scotland, say, the fortunate may perhaps witness a gathering of blackcocks at break of day early in the breeding season. The birds assemble in some open spot and indulge in the wild whirring calls that form their song of love and war, and the racket may be heard two miles off.

Then the tournament begins. It may be mere skirmishing, a display of fencing, or "sparring," or, as sometimes happens, these harmless encounters may develop into fierce fights and sometimes a duel to the death.

"At intervals during each separate fight, blackcocks emit a curious call; it is almost a hoarse screech, resembling the noise too painfully familiar to us, namely, that of cats on housetops supplemented by the said animals being afflicted with sore throats. The sound is both wild and unmusical in the extreme.

"We will suppose that the observer has come early on the scene, before the greyhens have made their appearance. The approach of one of the latter is the signal for an immediate cessation of hostilities on all sides, and intense excitement prevails amongst the assembled blackcocks. Her approach has been observed by a

From " Wild Life in the Tree Tops," by Captain C. W. R. Knight, F.R.P.S.

A LITTLE OWL IN ITS NESTING HOLE

This small and amusing species has become common in parts of England owing to artificial introduction from the Continent.   It is more frequently to be seen in the daytime than some of its relatives.

NEWTON'S BOWER-BIRD

There are several species of Bower-bird, all displaying remarkable habits in their courtship. In the case of the Gardener Bower-bird the nest itself is in the tree; at the foot of the tree a kind of cabin is built. "In front of it is arranged a bed of verdant moss, bedecked with blossoms and berries of the brightest colour"—regularly renewed as they wither. "The use of the hut, it appears, is solely to serve the purpose of a playing ground or as a place wherein to pay court to the female, since it is built long before the nest is begun."

single bird, who has been sharper than the rest in detecting the lady afar off . . . he will suddenly draw himself up to a rigid position of attention, till he is sure she is really coming. Having settled this in his mind to his own satisfaction, he throws himself into the air and flutters up a few feet, uttering the while hoarse notes with all the power and effect he can muster.

"This is, of course, done to impress the lady in his favour, and arouse in her breast a proper sense of admiration, which he considers his due. His example is immediately followed by all the others, who, on alighting, dance about in the most absurd manner, each one trying to see who can screech the loudest and be the most ridiculous in his antics.

"When a hen has alighted on the playing ground the male that is nearest to her pairs with her, and fights off any other that disputes his possession. She then meanwhile walks sedately round her lord and master, picking about the grass coquettishly, and pretending to be feeding. Each hen on arrival causes the same general excitement and is appropriated by one or other of the successful cocks till the harems are filled up, one cock having at times as many as six or seven hens. As the season advances, after the first few mornings of the hens coming to the ground, they resort to the same spot each day and stay with the same cock who has previously trodden them, and are not interfered with afterwards by other cocks, who acknowledge the superior claims of the male to whom they rightfully belong." [1]

In some cases there are special aids to display, such as the pouch in the neck of the Great Bustard, which the cock can distend at will and use as an aid in the erection of his feathers; Pigeons, too, have a similar habit of inflating their "crops," although they lack special plumes; and the Frigate-bird has an external pouch which itself serves as an ornament, being of naked skin, bright red in colour, and very extensible.

[1] J. W. Millais.

The Bower-birds of Australasia

Examples could be multiplied almost indefinitely, but we must here confine ourselves to one other case which has a novel feature of its own. The different species of Bower-birds found in Australasia build various types of "bowers" which serve as playgrounds in which the cocks court their mates. These "bowers" are often large and complex structures of twigs or flower-stems and are decorated with collections of blossoms, shells, or brightly coloured berries. One species builds a little cabin, some two feet high and three feet in diameter, at the foot of a tree and with a wide mossy "lawn" in front, while another makes a tunnel several feet long and completely roofed over with twigs. These bowers form the birds' courting grounds and are quite distinct from the nests, which are built in trees at a later stage.

Fighting with rivals plays a part of varying magnitude in the loves of different birds. Some species are well known for their pugnacity, the familiar Robin for instance; and in cock-fighting this has been turned to account as a source of human entertainment. In the domestic cock and in pheasants the development of spurs as weapons of offence is well known, and in some kinds of birds there are several pairs. Other birds fight with their wings, and lapwings may be seen buffeting each other in mid-air; an Egyptian relative of the Lapwing, the Spur-winged Plover, has a weapon on its wings which is said to make a fatal result no uncommon occurrence. The Ruff, a kind of sandpiper now numbered among the rarer English birds, has a frill of feathers round the neck, which is a shield of defence as well as an ornament for display in the regular tournaments which are held; the females, called reeves, lack the distinctive adornment.

Voice

The seat of the voice in mammals is in the larynx, at the top of the windpipe. In birds, however, the vocal cords are at the foot of the windpipe in a special enlargement called the song-box or syrinx. The sounds are due to the rapid passage of the air over the tense cords. In the course of evolution the significance of the voice has broadened out. From a simple parental call it became a means of recognition of any kindred, and in the course of ages it became expressive of particular emotions—emotions of joy and of fear, of jealousy and of content. While a certain amount of vocal ability is part of the hereditary make-up, there seems little doubt that the gift requires educating. The song of the first year is sometimes what one might call tentative and generalised. It improves with practice and is probably helped by emulation and imitation. The way in which some birds, e.g., skylarks, steal snatches of one another's music, suggests the importance of imitation as a factor in educating the vocal powers.

Song

We have spoken of song as the vocal part in the display of courtship, but it would be wrong to think of it as being no more. Song is, indeed, not confined to the breeding season, but the periods differ with the species; the extent to which the females can sing also varies. It is not possible to draw a sharp dividing line between true song and the notes which constitute the ordinary language of birds, and this gives another reason for not over-emphasising the sexual significance of song.

The definition of song must not be too strictly confined to notes which sound musical to human ears. Outside the ordinary song-bird group, there is quite commonly found some note or cry which is especially associated with the breeding season and which may be regarded as the equivalent of a song. Many of these cries seem harsh and discordant to us, but others have an obvious charm, at any rate, in their native surroundings; amid the rugged beauty of a wild moorland the weird bubbling spring-call of the Curlew is perhaps more appropriate music than the dainty lilt of the sweetest warbler. There are other notes, too, which are not vocal: pigeons, for instance, can clap their wings loudly together in flight, the White Stork rattles the halves of his beak like casta-nets, and the Snipe "bleats" or "drums" in springtime, as we have already remarked.

§ 13

NESTING HABITS

Primitive Nesting Habits

If the earliest birds were arboreal, as we have reasons for believing, the primitive nesting sites were doubtless also in trees. The elaborate structures made by many present-day birds, how-ever, are obviously products of a highly specialised habit which has been evolved in the course of ages. At an earlier stage the eggs would be laid in such natural sites as were available with-out the necessity of building, and modern examples of a similar habit are not wanting. A species of White Tern, for instance, inhabits tropical islands and frequently deposits its single egg on the strong horizontal leaf of a palm-tree. As Dr. H. O. Forbes says, "The egg is laid in the narrow angular gap between two leaflets on the summit of the arch of the leaf, where it rests securely, without a scrap of nest . . . yet defying the heaving and twisting of the leaves in the strongest winds. The leaf, as in all palms, goes on drooping further and further till it falls, and among the settlers [on Cocos Keeling Island] it is a subject of keen betting, when they see a tern sitting on an ominously withered leaf, whether the young bird will be hatched or not before the leaf falls. The result . . . has always been in favour of the bird; if the leaf falls in the afternoon, the tern will have escaped from the egg in the morning!"

Examples of birds which nest in holes in trees, in accordance with the probably ancestral custom, are the Owls, the Parrots, the Titmice, and of course the Woodpeckers.

Photo: "*Country Life.*"

EGGS OF THE GUILLEMOT

These show an extraordinary range of variability, both in marking and in colour. Browns, yellows, reds, greens, blues, and whites are found, and they may be blotched, spotted, lined, or streaked. The shape is believed to serve a useful purpose in preventing the egg from rolling—except in a small circle—and thus from falling off the cliff ledge on which it is laid.

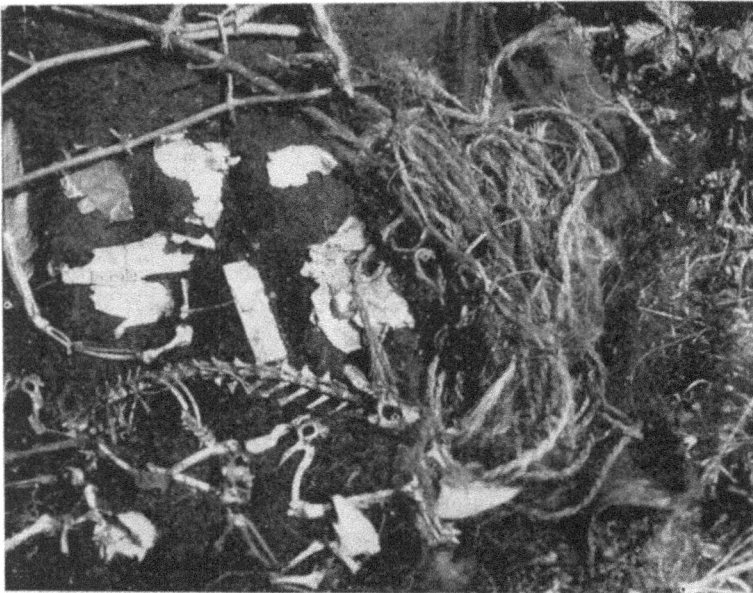

From "*Wild Life in the Tree Tops,*" *by Captain C. W. R. Knight, F.R.P.S.*

STRANGE NEST MATERIALS OF A CARRION-CROW

Some nests are composed of the bones of departed birds and mammals. There may be **also** quantities of string, cigarette packets, and even, in one case, a lady's handkerchief.

Photo: British Museum (Natural History).

THE WONDERFUL NEST OF REMEZIA, THE PENDULINE TIT-MOUSE

Note the funnel-shaped opening leading to the chamber; below is a pocket which is supposed by some to be a roosting-place for the male bird. The nest is made of cotton and seed down, and to the touch resembles a fine felt carpet.

Photo: J. H. Symonds.

THE REED-WABBLER'S NEST IS A BEAUTIFUL CUP-SHAPED STRUCTURE SUSPENDED AMONG THE REEDS

The Imprisonment of the Hornbill

Another hole-nester is the Hornbill, of which various species are found in many tropical lands, and its story is a very strange one indeed. When the eggs are laid and the hen begins to sit, the opening in the tree-trunk is walled up with mud by the cock until only a small orifice remains through which the sitting bird can put no more than her head. The device is doubtless a means of defence against snakes or other enemies, but it involves the imprisonment of the hen during the whole period of incubation. During this time, however, she is by no means left to starve, but is fed assiduously through the "grille" by her devoted mate, who is indeed said to work so hard and to forage so unselfishly that he is worn to a mere shadow of his former self before the task is done.

Among the tree-nesting birds the most primitive type of wholly artificial nest seems to be the platform of sticks or twigs made by such birds as eagles, herons, and pigeons. These structures are often of great size, being added to year after year. The simplest platforms are quite flat, but others are more or less cup-shaped, as in the case of crows. Finally, this type reaches its highest point in those birds which add a dome-shaped roof.

More Complex Structures

More promising material is used by most of the small birds which nest in trees or bushes, and with pliable twigs, grasses and roots, moss, and perhaps animal hair, much more complex structures are possible. The Finches, for example, make elaborate and beautiful cup-shaped nests, while others, such as the Wren and the Dipper, make spherical nests which can be entered only by a small hole in one side. In addition to the actual structure there is often a distinct lining of specially selected material; for this purpose small feathers, hair, and fine fibres are greatly favoured, but in the familiar case of the Song-Thrush, for instance, a complete lining of hardened mud is a characteristic feature. Few nests reach such a high development as that of the Tailor-bird of India, so called from its habit of "sewing" leaves together to make a beautiful pouch, a very triumph of the nest-builder's art.

Burrowers

From nesting in holes in trees to nesting in holes in the ground is an easy transition, and the gap is bridged by birds like the Stock-Dove, which use either site according to the opportunities which a particular district may happen to afford; this bird gets its name from the habit of nesting in holes in the "stocks" of old trees, but among the sand-dunes on many parts of the British coastline it uses rabbit-burrows instead. In similar haunts we may also find another burrow-nester—the bird which Mr. W. H. Hudson calls "the strange and beautiful Sheldrake." Unlike most of the duck family the male Sheldrake is not subject to an "eclipse" moult in the midst of the breeding season, and he is therefore able to stand by his mate, who, furthermore, has a bright plumage similar to his own.

Other birds which nest in burrows are the Petrels, some Penguins, the Kingfisher, and the Sand-Martin. The last-named nests in colonies, and each pair tunnels many feet into the chosen bank and hollows a little chamber at the end; the Bee-eater makes a similar tunnel, which may be as much as ten feet long. As with holes in trees, a lining may be added, say of grass or other vegetation; the Shelduck, like others of its kind, uses a plentiful supply of down plucked from its own breast, while the Kingfisher lines its nest with an unsavory collection of fishbones and other remains of its prey. The Megapodes go to the extreme of completely burying their eggs either in pits or under specially constructed mounds.

Ground-Nesting

Very many other birds nest either on the open ground or among the long grass and herbage. Sometimes there is a well-built nest among grass, as in the case of the Skylark or the Meadow-Pipit; at other times there may be a bulky heap of vegetation or of other material; the Cormorant, for instance, often raises a mound of seaweed, and some kinds of Penguin build a Spartan nest of stones. Still, again, there may be a mere hollow scraped in the ground, as in the case of the Lapwing or of the Tern, perhaps with a lining, a pretence at a lining, or with no lining at all. Finally, the bird may lay its eggs on the ground without any attempt at a nest, as the Oyster-Catcher does among the riverside shingle.

Cliff-Nesting

Somewhere between the tree-nesters and the ground-nesters we must place those birds which nest on cliffs, for although a nest on a rock ledge may seem in some ways very like a nest on flat ground, the dependence on inaccessibility rather than on concealment makes the habit also akin to tree-nesting. Some of the burrowers, like the Puffins and the Petrels, might well be classed in this group as their holes are usually on precipitous faces, but more typical are those species which breed on the open ledges, like the Guillemot and the Razorbill. A highly specialised type of nest, too, is that which is built of mud against the sheer rock face, and for this purpose—as in the House-Martin—the habitations of man are often found to serve as well as natural faces of rock. Sometimes the mud and other materials are made more coherent by the addition of the salivary secretion of the builders, and with the Edible Swift of Borneo this substance, like hardened glue, forms practically the whole structure and is the source of the "bird's-nest soup" beloved of the Chinese gourmet.

The Use of Old Nests

Many birds return to their old nests and use them again and again, while other kinds habitually build afresh each year

There are birds, too, which commonly use the old nests of other species, with or without additions of their own, although they are not always incapable of building for themselves if faced with the necessity. This habit is not uncommon in the case of birds of prey; the Kestrel, for example, often uses the old nests of crows and pigeons. The Green Sandpiper, belonging to a very different order of birds, uses the old nests of thrushes and other tree-nesting birds—and even squirrel's "dreys"!—although most of its own kin are typical ground-nesters.

§ 14

Chicks and Nestlings

It is impossible to leave the main question of nesting habits without some reference to the striking differences observable among the newly hatched young of birds. These fall into two well-marked groups in accordance with the condition and stage of development at the date of leaving the egg. Technically these groups are the *nidifugous* and the *nidicolous,* terms which we may translate as nest-quitting and nest-dwelling, though perhaps something of the distinction is conveyed in the two ordinary names "chick" and "nestling." The chick of the domestic fowl is notoriously a nest-quitter; so also are ducklings, whether domestic or belonging to one of the many wild species, and so likewise the young of the plover kind. All these birds leave the egg prepared to take an immediate active part in life; they are open-eyed and lively, able to walk—and, in appropriate cases, to swim—and capable of finding their own food with no more than the guidance and protection of the parent. Contrast these with, say, young thrushes—helpless, blind, almost naked, and rather repulsive-looking creatures, which would die miserably without the food their parents so assiduously bring. The difference is, indeed, a most striking one, but some of the nest-dwelling young are not quite so unlike the more active chicks; the nestlings of the birds of prey and of the owls, for instance, are clothed in

NESTS OF THE EDIBLE SWIFT

The saucer-shaped nests are made entirely from the saliva of the birds, and form the source of the "birds'-nest soup" which is considered a delicacy in China.

GUILLEMOTS

The Guillemot, one of the Auks, is common on precipitous parts of the British coastline during the breeding season. A single egg is laid on the bare rock of a narrow ledge. At other times of the year the birds keep to the sea, but dead ones are often washed up on the beach after storms.

down and are open-eyed and alert, although they remain in the nest at first and are wholly dependent on their parents for food.

Transporting the Young

We have an illustration of how some birds make use of their wits in the way they transport their young. In this connection Lord Grey recently told how he watched a Wood-Duck (Carolina) whose nest was a hole in a tree 21 feet from the ground and 300 yards from the water. "Presently the duck flew down from the hole into the grass, and began calling; then one by one the little ducklings came to the edge of the hole and *fell to the ground*. When measured the nest was found to be 2 feet below the hole. For the newly hatched birds to climb that distance, to fall 21 feet, and then follow their mother 300 yards to the water was, I think, a tremendous tribute to the energy of nature."

The female woodcock, when threatened with danger, is known to transport her young, one at a time, to another place. She does so by carrying the young ones with her feet, holding them in her claws, or pressed between her thighs. It is also said that where she nests at a distance from the feeding-ground, she will carry her young to and fro in the morning and evening.

§ 15

The Study of Birds' Eggs

We cannot here discuss fully the eggs of birds. A wealth of matter for speculation lies in the why and wherefore of size and shape, of texture and colour, and of the numbers forming a clutch. All these characters show wide limits of difference, but on the whole they remain very constant and characteristic for any one species.

Size and Shape of Eggs

The size of the individual egg is variable, apart from the question of due proportion to the size of the parent bird concerned; this is related in a large degree to the length of the incubation period, while this in turn depends to an important extent on the state of development of the young when hatched, a subject which has already been discussed.

In texture of shell, eggs vary from the brilliantly polished egg of the Tinamous to the soft chalky eggs of the Cormorant, from which the white outer surface can be scraped to show a pale-blue layer beneath. Thickness of shell is also a variable factor, apart from the mere relation to general size.

Egg Coloration

It is, however, the colour of eggs that have always attracted most attention; some of these are exceedingly beautiful both in tint and in the patterns of marking. Blues and greens are common especially among tree-nesting birds, while ground-nesters usually show neutral brown tones which are most effective for purposes of "camouflage"; some splendid red tones are characteristic of the birds of prey. Markings may be small spots or larger blotches, and they may be evenly distributed or concentrated in a particular zone; fine lines also are found in some cases, witness the Buntings, and in many birds there is a plain marked ground-colour. Pure-white eggs are usually found in species which nest in holes, and this is perhaps of some use in the dark, although the more important point is probably the absence of any occasion for an attempt at "camouflage" coloration. Coloration in many instances serves a protective purpose, and, generally speaking, it is related to some extent to the nature of the bird's environment. There are, curiously, no pure black eggs.

§ 16

Behaviour of Birds

More than any other creatures, birds have claimed the attention of those who are fond of what Fabre called "scrutinising life." There is often an extraordinary subtlety as well as beauty in their habits. They are big-brained animals, and the senses of sight and hearing are developed to great perfection.

The question is how much in the behaviour of birds we must ascribe to instinctive endowment, that is, to inborn impulsions or hereditary nervous predispositions, and to what extent must we credit the bird with intelligent learning? When a young moorhen swims deftly the first time it touches the water, or dives perfectly when the fit and proper stimulus is forthcoming, we interpret this as instinctive. Its physiological side is a concatenation of reflex actions. Its psychological side is inborn impulse and endeavour. Similarly, when an unhatched lapwing utters its characteristic call-note "peewit" from within the egg, we say this is instinctive—independent of instruction, learning, or imitation.

But a different note is sounded in the behaviour of the Greek eagle, which lets the tortoise fall on the rocks from a great height, so that the carapace is broken, or in the similar device of the Rook that lifts the freshwater mussel and lets it fall on the riverside stones. The Herring-Gull sometimes lifts the sea-urchin, or the clam, in its bill, and lets it fall on the shingle, so that the shells are broken. Without necessarily supposing that these birds thought out the expedient, we can hardly avoid the conclusion that they utilise the discovery intelligently. In many cases the bird must be credited with an appreciation of circumstances, with an awareness of what is significant, and with a capacity for learning. The young chick's capacity for rapidly learning simple lessons, mostly associations, has been proved up to the hilt by many experiments.

"In the quiet of the wood one sometimes hears the song-thrush breaking snail shells on its stone anvil, and one may easily find the telltale evidences of its appetite. Is this habit, which comes so near using a tool, an inborn gift or has it to be learned? The answer is given by Miss Frances Pitt in her admirable *Wild Creatures of Garden and Hedgerow*. To a young thrush which she has brought up by hand she offered some wood-snails (*Helix nemoralis*), but he took no interest in them until one put out its head and began to move about. The bird then pecked at its horns, but was bewildered when the snail retreated within the shelter of its shell. This happened over and over again, the bird's inquisitiveness increasing day by day. The thrush often picked one up by the lip, but no real progress was made till the sixth day, when the thrush beat a snail on the ground as it would a big earthworm. At last on the same day he picked up a shell and hit it repeatedly against a stone. He tried one snail's shell after another, until after fifteen minutes' hard work he managed to break one. After that all was easy. He had cracked his first snail. After long trying he had found out how to deal with a difficult situation. We may say, then, that while a certain predisposition to beat things is doubtless inborn, the use of the anvil is no outcome of a specialised instinct, it is an intelligent acquisition."

The general impression that one gets in regard to the cleverness of birds in such activities as nest-building, capturing booty, and dealing with food is that on an instinctive basis, varying in definiteness, there is built up a superstructure partly due to easy education and subsequent imitation, and partly due to an intelligent appreciation of the lessons of experience. But an appreciation of the relative importance of "nature" and "nurture" requires careful observation and experiment.

BIBLIOGRAPHY

BEEBE, *The Bird* (1907).
CLARKE, *Studies in Bird Migration* (1912).
HEADLEY, *The Flight of Birds* (1912).
HUDSON, *British Birds.*
KIRKMAN and others, *The British Bird Book* (1911-13).
MATHEWS, *Field Book of Wild Birds and Their Music* (2nd ed., 1921).
NEWTON, *A Dictionary of Birds* (1896).

PYCRAFT, *A History of Birds* (1910).
SAUNDERS, *Manual of British Birds* (2nd ed., 1899).
SHARPE, *Wonders of the Bird World* (1898).
THOMSON, *British Birds and their Nests* (1910).
WITHERBY and others, *A Practical Handbook of British Birds* (1920-2).

XIII
NATURAL HISTORY

.

NATURAL HISTORY

II. MAMMALS

IN an earlier chapter we have dealt with the evolution of animals in general, their haunts or habitats, their everyday functions, their behaviour, and what we have called the dawn of mind. Here we select one class, that of Mammals, and, presupposing what has gone before, we shall discuss them in the main from one point of view—how they are suited to the particular conditions of their life.

Origin of Mammals

The genealogical tree of animals splits at the top into Birds and Mammals, and these are on quite different lines of evolution. They are not related to one another, except to this extent, that they have a common ancestry among the extinct Reptiles, as we have already seen. For just as Birds sprang from some uncertain stock of bipedal Dinosaurs, so Mammals must be traced back to another extinct Reptilian stock—the Cynodonts. These Cynodonts (also known as Therapsids) occur as Triassic fossils in Africa and North America; though they were genuine reptiles they had very mammal-like skulls (see figure facing p. 454). Thus the teeth may be distinguished as incisors, canines, and molars, as in a dog; hence the name Cynodont, "dog-toothed."

The Earliest Mammals

The earliest mammals were small creatures, the largest no bigger than a rat. The teeth of some of them indicate insect-eating, the teeth of others point to a herbivorous habit. The sharp incisors of some types may have been used to pierce the shells of the eggs of waning Dinosaurs. According to some authorities, many of the early mammals were arboreal, denizens, perhaps, of estuarine and swampy forests. The advantage of such a habitat or mode of life is suggested by the scant vegetation of the arid ground.

Archaic Mammals

During the geological Middle Ages (Mesozoic) the mammals did not make much headway. Their opportunity was not yet; it was still the Age of Reptiles. The mammals continued a little folk, probably for the most arboreal, keeping out of the way of the huge carnivorous dinosaurs, "stalking terrors such as the world never saw before nor since."

As a matter of fact, however, the giants disappeared, and the pigmies had their innings. With the dawn of the Tertiary time, the mammals began to possess the earth. Their giant enemies had gone, and it is probable that the vegetative conditions became more favourable. The grass began to spread like a garment over the earth.

Modern Types of Mammals

Progress was at first very gradual; the early Tertiary mammals were still pigmies and with very small brains; but the point is that they began to radiate out into old-fashioned marsupials, carnivores, and hoofed mammals—some of the last attaining elephantine dimensions.

As the primitive and archaic mammals disappeared, there rose up in their stead the mammals of the modern type—with better brains and more plastic feet and teeth. We refer to such families as Cats, Horses, Elephants, and Monkeys. The original headquarters were probably in some northern or circumpolar land, which enjoyed a warm and equable climate.

§ 1

The Egg-laying Mammals

There are three strange Australian mammals that occupy a position quite by themselves—the Duckmole (Ornithorhynchus), the Spiny Ant-eater (Echidna), and another ant-eater (Proechidna). They differ from all other mammals inasmuch as *they lay eggs,* thereby harking back to the habit of many reptiles. In the form of their shoulder-girdle, in their relatively large eggs with much yolk, in their very variable temperature, and in many other ways they betray their affinity with reptiles, and they must be regarded as very primitive mammals persisting from ancient days.

The Duckmole, or Duck-billed Platypus (18-20 inches in length) lives beside lakes and streams, and grubs at the bottom or among water-weed for small animals, which it collects in cheek-pouches and chews at leisure with its eight horny tooth-plates. For its true teeth do not last for more than a year. Its fore-feet are webbed, and it is a clever swimmer and diver. But the feet are also clawed, and the quaint creature makes a long burrow in the bank, with two openings, one above and one under the water. The jaws are flattened like the bill of a duck and covered with soft sensitive skin, expanded into a flexible collar where the bill joins the rest of the skull. The eyes are small; the ear-holes are closed by a flap; the tail is strong and helps in swimming; the brownish fur is short and soft; the animal can roll itself up into a living ball, and sleeps in this attitude. In the recesses of the burrow two eggs are laid, each about three-quarters of an inch long, enclosed in a flexible white shell, through which the young one has to break its way. There are no teats or mammæ for the young one to suck, and the milk simply oozes out by numerous pores on a bare patch of skin on the ventral surface of the mother. It is licked up by the offspring—a very primitive arrangement.

The Spiny Ant-eaters live in rocky regions and burrow

rapidly with very strong claws.   They seem almost to sink into
the ground.   When they get among rough herbage they take
firm hold with their feet and are very difficult to dislodge.   The
snout is prolonged into a slender tube, through which a mobile,
sticky, worm-like tongue is protruded on the ants which form
the staple food.   No traces of teeth are to be seen, even in the
embryo.   As in the Duckmole, the male has a well-developed
spur on the hind-leg, perforated by the duct of a gland, but its
use is obscure.   The egg seems to be placed by the mother in a
temporarily developed pouch, which is said to be comparable to
a greatly enlarged teat of the type seen in the cow.   Within
this pouch the milk oozes out.   There are no stranger animals
in existence than the Duckmole and the Spiny Ant-eaters.   They
might almost be called "living fossils."

§ 2

The Pouch-bearing Mammals

The second grade among present-day mammals is that of
the Marsupials, which are now confined to Australia except in
the case of two families—the American Opossums and Selvas.
In most cases the female has a pouch or marsupium developed
around the mammæ, and in this pocket the prematurely born
young are stowed away and carried about till they are able to
fend for themselves.   In many opossums the pouch is absent,
and the mother carries the young ones on her back, with their
tails coiled round hers—a quaint device.   In marsupials in general, the young ones are born very helpless, unable even to suck.
The mother takes her young one in her mouth, and puts it into
her skin-pouch, within which lie the teats or mammæ.   The
mother adjusts matters so that the mouth of the young one closes
on a teat, which then swells a little, and, as the prematurely born
offspring cannot suck, she injects the milk down the gullet by
contracting a special musculature.   The milk might "go down
the wrong way" and choke the offspring, were it not that the

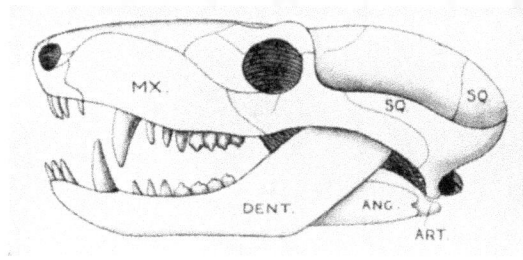

SKULL OF A CYNODONT EXTINCT REPTILE. (*After Broom.*)

The Cynodonts do not seem to have been very far from the direct ancestors of Mammals. The arrangement of the teeth as incisors, canines, and molars is very mammal-like. But the lower jaw remains a complex of several bones, whereas in mammals there is but one bone on each side. Some of the bones are named: MX. maxilla; SQ. squamosal; DENT. dentary; ANG. angular; ART. articular.

Photo: W. S. Berridge.

ECHIDNA OR SPINY ANT-EATER

This primitive egg-laying mammal ranges from Australia through the Papuan region. There is a related genus, Prœchidna, in New Guinea. The body is covered with strong spines mingled with hair. There are three massive claws suited for burrowing. The mouth is absolutely toothless; the tongue is worm-like; the food consists mainly of ants and other insects. When the egg is laid, the mother takes it in her mouth and places it in her pouch. The shell is broken by the emerging young one. After a time the mother removes the young one from the pocket and leaves it in a burrow while she hunts at night. But she restores it for feeding purposes. The cerebral hemispheres are well convoluted, and the creature is not stupid. It is imperfectly warm-blooded, and hibernates.

	ELEPHANTS	APES	LAND CARNIVORES	
	UNGULATES	MONKEYS	AQUATIC CARNIVORES	
CETACEANS	RODENTS	LEMURS	INSECTIVORES	BATS
	SIRENIA		EDENTATA	
	MARSUPIALS			
	MONOTREMES			

GROUPING OF THE ORDERS OF LIVING MAMMALS

At the bottom of the scale are the primitive, egg-laying Monotremes, represented by the Duckmole and the Spiny Ant-eater. Above them are the pouched Marsupials, of which the Kangaroo is a type. The remaining Orders fall within the group of Placentals, in which there is an intimate connection between mother and unborn young. Of these, the Edentates (e.g. the Sloth) and the Sirenians or "Sea-Cows" may be ranked as archaic forms. Higher up we have the Insectivores (e.g. the Hedgehog); the aquatic Carnivores (e.g. Seals) and the terrestrial Carnivores (e.g. Lion); the Rodents (e.g. the Rat); the Ungulates or hoofed quadrupeds, and the Elephants; and, marking extremes of mammalian life, the winged Bats and the open-sea Cetaceans or Whales. Finally, there is the great Monkey stock, including the old-fashioned Lemurs, the true Monkeys, and, highest of all, the Anthropoid Apes.

OPOSSUM AND YOUNG

In some of the Opossums, especially the smaller species, the mother carries the young ones on her back with their tails twisted round hers. Azara's Opossum may carry eleven and yet climb quickly. The habit may occur even when there is a pouch, but that structure is generally absent in the Opossum family (Didelphyidæ). The Opossums are also peculiar in being confined to North and South America.

glottis (the entrance to the windpipe) is shunted forward in the young creature so as to press against the posterior nostrils at the back of the mouth. Thus breathing goes on undisturbed by the injection of milk. A similar adaptation is seen in the Baleen Whale—another mammal—when it is rushing through the water with its great mouth agape, and also in the Crocodile—a reptile, not a mammal—when it is drowning its prey.

Playing 'Possum

As an individual example of a marsupial we may take the Opossum (Didelphys)—which Mr. Ingersoll calls a "grey, grunting, snarling, pilfering, dunder-headed, and motherly creature." It is not a good type, for it is American, not Australian, and in most of the species of the genus the pouch is conspicuous by its absence. But the Opossum was the first marsupial to be known to the civilised world. Opossums are mainly arboreal and insectivorous, but there is considerable variety of habitat and diet. They are notorious for playing 'possum, and we wish to incorporate what Ingersoll says of a pouched species in regard to this puzzling "ruse" in his *Wit of the Wild* (1921). A mother opossum will face up to an enemy that threatens her half-grown young, and male opossums will fight to the death at the courting time. So the creature does not lack courage! If it detects danger in advance—and every hand is against it—it will hasten up a tree and hide. So the creature does not lack discretion. "In other cases—just what or when it would be hard to define exactly, but apparently in the presence of something so large as to make resistance idle—the animal, when attacked or cornered, will fall limp and 'dead.' You may roll the creature about with your foot, explore the pouch, pick it up and carry it by its tail, offer it almost any indignity, and it will, in most cases, neither resist nor complain; but take your eyes off it as it lies upon the ground, and it will soon jump up and scuttle away, or if you pick it up carelessly enough to give it a chance it may nip you savagely." But the question is inevitable: "Of what service is the ruse?" Would the carnivore or the bird of prey that liked opossum flesh—dogs won't touch it—care whether the creature is dead or pretending to be dead? Mr. Ingersoll's ingenious suggestion is that playing 'possum is an instinct that arose in the geological Middle Ages in relation to the dull-witted big reptiles—as a rule, land reptiles do not feed on carrion—and that it persists nowadays as an anachronism in circumstances where it is oftener fatal than protective.

§ 3

The Placental Mammals

The third grade of modern mammals includes the carnivores, the hoofed animals, the monkeys, and so on—to all of which the term "placental" is applied. In adaptation to the difficulties of terrestrial life, there has been an evolution of viviparous arrangements. The Monotremes, as we have seen, lay eggs; the Marsupials bring forth their young prematurely; the Placentals have established a more or less prolonged ante-natal partnership between the mother and the unborn young. The linking structure between the two is the placenta, which brings some of the blood-vessels of the unborn young (or fœtus) into close contact, although not union, with the blood-vessels in the wall of the mother's womb (or uterus). No solid particle, unless it be a living microbe or a wandering white blood-corpuscle, can pass from the mother to the offspring, but there is a transfusion of fluid and gaseous material between the two partners. What does the offspring get from its mother? Dissolved nutritive material, oxygen, water, salts, and some subtle chemical messengers called "hormones." What does the offspring give to the mother? Dissolved waste-materials, carbon-dioxide, watery fluid, and again some "hormones." The mother gives much and gets little; but it seems justifiable to say that the internal secretions or hormones contributed by the unborn offspring to the mother assist

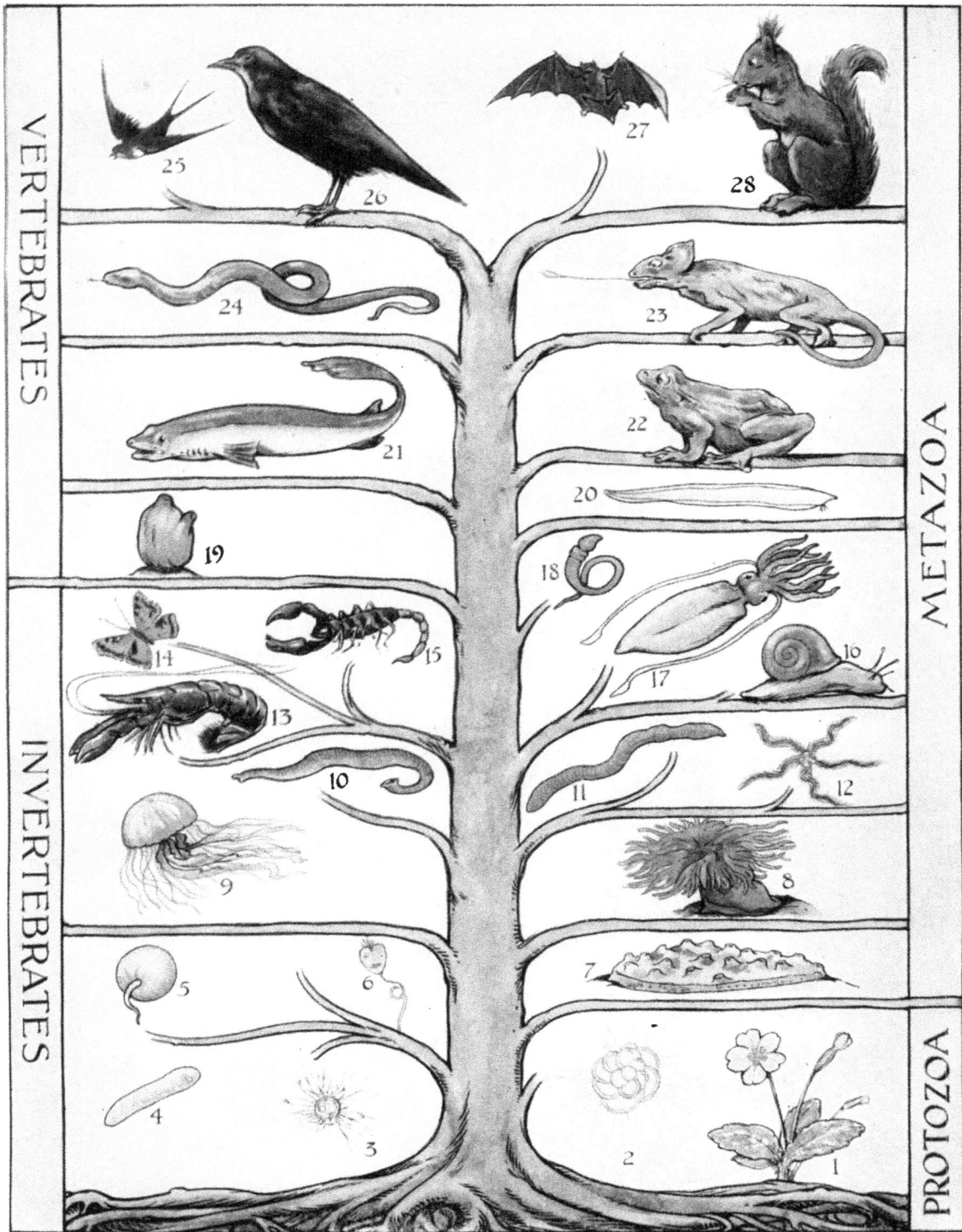

PICTORIAL REPRESENTATION OF THE GENEALOGICAL TREE OF ANIMALS

1. A plant, indicative of the Vegetable Kingdom—on another line of evolution.—2 and 3. Chalk-forming animals or Foraminifers.
—4. A parasitic Gregarine.—5. The Night-light Infusorian, Noctiluca.—6. The Bell-Animalcule, Vorticella. All these are Protozoa, unicellular animals. Multicellular animals are called Metazoa.—7. Crumb-of-Bread Sponge.—8. A Sea-Anemone, and 9. A Jelly-fish —both Cœlentera or Stinging Animals.—10. A Leech, and 11. An Earthworm, both Annelids.—12. A Brittle-Star, representing Echino-derms.—13. A Lobster, representing Crustacea.—14. A Butterfly, representing Insects.—15. A Scorpion, representing Arachnids.—16. A Snail, and 17. A Cuttlefish, both representing Molluscs.—18. Balanoglossus, a worm-like type intermediate between Invertebrates and Vertebrates.

Among Vertebrates: 19. A Sea-Squirt or Tunicate.—20. A Lancelet.—21. A Shark (Fishes).—22. A Frog (Amphibians).—23. A Chameleon, representing the Lizard order of Reptiles.—24. A Snake, another type of Reptile.—25. Swallow, and 26. Rook, represent-ing Birds.—27. Bat, and 28. Squirrel, representing Mammals.

in her health and enable her to make the most of her food. Before the young one is born, chemical messengers have been carried by the blood to the mother's mammary glands, so that they are stimulated to begin the production of milk. There is much of this physiological telegraphy in the business of living.

It is probable that the long-drawn-out ante-natal development has greatly favoured the improvement of the brain. Thus everyone knows how wide-awake a foal is after its long sleep of eleven months within its mother's womb. But it must be added that the structure of the brain in placental mammals had got on to lines much more promising than in marsupials. Granting this, we seem justified in saying that the prolonged gestation, plainly adapted to the exigencies of terrestrial life, opened up the possibility of being born with an advanced brain equipment. In the same way, the prolonged infancy, familiar in mankind, has its great rewards as well as its great risks.

It is interesting that mammals should bear a name that emphasises the mother's breasts, and this strikes a true biological note. For the success of mammals is wrapped up with their maternal care, taken in conjunction with improved brains. To the difficulties and limitations implied in the struggle for existence, some mammals have answered back by evolving teeth and horns, others by evolving swiftness, others by evolving armour, others by evolving wings—but the answer back that is common to them all is the maternal sacrifice and devotion.

§ 4

Many Habitats

Like the reptiles before them, mammals have sought out many habitats, and have become adapted to as many modes of life. Perhaps it was in the trees that they served their apprenticeship; in any case they have tried all possible haunts, entering every open door of opportunity. This is what Professor H. F. Osborn calls "adaptive radiation," and he distinguishes as many as twelve habitats. (1) There are terrestrial mammals, walking like the elephants, running like the antelopes, jumping like the kangaroos. (2) But some are burrowers as well as runners, as rabbits well illustrate. (3) Then there are thoroughgoing burrowers, like the moles, which have conquered the underground world. (4) Some are as much at home in water as on dry land; we think of the roving otter and the polar bear. (5) Perhaps a separate division may be made for those mammals that frequent streams, after the manner of beavers and the familiar water-vole —which can hardly be saved from its popular name of "water-rat." (6) The shore of the sea is the habitat of seal, sea-otter, and walrus. (7) The open-sea mammals are the cetaceans large and small, from whale-bone whale to porpoise. (8) Professor Osborn takes the deep-diving finback whales as examples of mammals that actually explore the great abysses, but this is perhaps stretching a point. (9) Then there are the betwixt-and-between mammals transitional between arboreal and terrestrial life, like the macaque monkeys and the gorilla. (10) Strictly arboreal types are well represented by squirrels, tree-sloths, and lemurs. (11) The volplaning "flying squirrels" and "flying phalangers" form another interesting betwixt-and-between group, essaying the conquest of the air in their daring parachuting from tree to tree. (12) Finally, the bats are true fliers— aerial mammals.

It is useful to recognise this variety of habitat, for it shows how diverse the life of mammals must be, and the impression of diversity grows when we remember that in most habitats there are several distinct possibilities of food-getting. Thus a mole is a carnivorous burrower, while a vole is a vegetarian burrower; a small bat is an insectivorous flying mammal, while a big bat is usually a fruit-eater. It is very interesting to find that almost every haunt and diet illustrated by mammals has also been utilised by reptiles, either living or extinct. This suggests that evolution has proceeded on an ever-ascending spiral.

Birds and mammals have evolved, as we have already said, on entirely different tacks, but it is not unprofitable to notice that they have often made the same kinds of experiments. The Ostrich is a running bird, the Antelope a running mammal; the Owl is a nocturnal bird, the Hedgehog is a nocturnal mammal; the Storm-Petrel is an open-sea bird, the Dolphin an open-sea mammal; the Sand-Martin is a burrowing bird, the Mole a burrowing mammal, and so on. For a long time there were no flying mammals to vie with the flying birds; but eventually there was the evolution of bats, doubtless from an arboreal insectivorous stock.

Aquatic Mammals

It is instructive to consider some of the thousand and one ways in which mammals are specially adapted to the various haunts and conditions in which they live. But only a few illustrations can be given, beginning with aquatic mammals. In whales the tail has been transformed into a propeller, which sculls the water first to one side and then to the other, and great speed is attained in swimming and diving. With these swimming powers is associated the almost worldwide distribution of many cetaceans, like the Sperm Whale and the Southern Right Whale. In seals the hind-limbs are bound up with the tail, a conjoint propeller which churns the water from side to side being the result. In the walrus the hind-limbs are helped by the great paddle-like fore-limbs, which are also used for clambering on the slippery ice. The Common Seal has a remarkable way of moving on land. It arches up its body, bringing the hind-limbs and tail towards the head, and then suddenly straightens itself away, thus jerking the body forwards. In swimming the Beaver uses its trowel-like flattened tail; the Duckmole has webbed fingers; the

Water-Shrew has special hairs on the sole and toes of its hind-foot, which are spread out like a comb in swimming, but become appressed when the little creature runs on land. The long tail of the Water-Shrew serves as a rudder; it is somewhat flattened vertically and bears a fringe of long hair on its ventral surface. The adaptations to aquatic life are many: thus there is often a reduction of friction by the disappearance of external ears, as in seal and whale; hair is almost quite gone in cetaceans, though those that remain about the mouth may be very useful in their exquisite tactility; the absence of hair, which normally serves as a non-conducting robe, is compensated for by the development of a layer of blubber—just an exaggeration of the deposit of fat (*panniculus adiposus*) which is formed under the skin in most mammals (the Common Hare a noteworthy exception); the mother whales have an arrangement for giving their baby a huge mouthful of milk at a gulp, for suckling cannot be very easy in the open sea. It is said that the Northern Right Whale may remain under water for an hour and twenty minutes, and in adaptation to this prolonged immersion there is a huge chest cavity, and also a development of wonderful networks (*retia mirabilia*) of arteries which store pure blood and keep the tissues oxygenated when respiration in the ordinary sense has come to a standstill. According to Lillie, a rorqual may remain eight to twelve hours under water, and it is possible that in this case a sort of skin-respiration (familiar in frogs, for instance) is effected by means of numerous very vascular longitudinal ridges on the underside of the rorqual's throat. Besides the positive fitnesses, of which some illustrations have been given, there are negative adaptations. Thus, in thoroughly aquatic mammals, such as whales, there can be no smelling, and the olfactory organ is naturally degenerate. For what is useless is rarely conserved. Or, again, the cetaceans, which have their eyes continually washed with water, have no third eyelid—which is used in other mammals, except man and monkeys, for cleaning the front of the eye.

YOUNG OF COMMON SEAL ON THE BEACH, SHETLAND

The Common Seal (*Phoca vitulina*) is a sociable animal, living in small herds where the conditions are suitable. On British coasts these safe places are rapidly disappearing, for the young, left on shore by the mothers when they go a-hunting, are often killed. The drawing gives a fine impression of the charming attitudes of the young creatures.

YOUNG OF BEARDED SEAL

As in many other cases, the young of the Bearded Seal (*Erignathus barbatus*) has a uniformly light-coloured coat—in this species practically white. It is retained for some weeks. The young of the Harp-Seals (*Phoca grœnlandica*) are also called "white-coats." When the young one is lying exposed on the ice there may be protective value, or, more probably, physiological comfort, in being white. The Bearded Seal is a North Atlantic species, occasionally visiting British shores.

Photo: F. R. Hinkins & Son.

COMMON MOLE (*Talpa europœa*).

One of the conquerors of the underworld, adapted to its subterranean life in its barrel-like shape, reduction of friction (e.g. no external ears), elongated muzzle, shovel-like hand, strong breast muscles, and powerful neck for tossing the earth. The hair has no "set"; the minute eye is well concealed. The mole is a representative of the dwindling order of Insectivora, but its range still extends from Mull to Japan.

§ 5

Subterranean Mammals

As life on the surface of the earth is attended by great risks, which have to be met by special adaptations, it is not surprising that many mammals should seek refuge underground or should combine terrestrial and subterranean habits. Of adaptations to thoroughgoing subterranean life the Mole is perhaps the finest instance. Its hand is turned into a strong shovel, with which it literally "swims" in the earth. To the inside of the thumb there is a special sickle bone, which broadens the digging surface. The breast muscles are like an athlete's, and those of the very short neck are well suited for tossing the earth. There are no projecting ear-trumpets, for these would be much in the way; the eye, unnecessary in darkness, is reduced to a pin-head size (1-25 of an inch in diameter), and is protected from injury by being well hidden in the hair of the head; the position of the nostril rather under the tip of the snout and a lip-fold in front of the mouth serve to keep the earth out; the hair of the body has no "set" and is easily kept clean, moreover it does not get disordered when the burrower moves backwards; the crowns of the back teeth are covered with sharp cusps, most admirably suited for crunching insect larvæ and the like. Truly the Mole is a bundle of adaptations. The Common Mole burrows in soft soil, and its hand is therefore broad; but the Cape Golden Mole and the quite unrelated "Marsupial Mole" burrow in hard soil, and their hands are very narrow, with a great strengthening of two of the fingers. This is plainly as it should be, and the impression of fitness grows when we consider details. Thus the Marsupial Mole, which presses its head into the earth, has its neck vertebræ solidified.

The Mole

We have mentioned the Mole's adaptation to subterranean life, but this extraordinarily interesting mammal claims further attention. It is not only a bundle of adaptations, it is an antiquity; it was long ago one of the discoverers of the under-world; it ranges successfully from Mull to Japan; it lives an unusually strenuous life; it has the charm of elusiveness and idiosyncrasies. It has four modes of locomotion. Ordinarily it "swims" deeply in the earth, using its hands to force the earth to either side, and scratching backwards with its hind-feet. It can burrow for a considerable distance without making a mole-hill. Secondly, when there is food, e.g. leather-jackets (the larvæ of the crane-fly or daddy-long-legs), to be got near the surface, the mole works along in a shallow groove, often breaking to the open, and leaving a discernible track. In this shallow burrow-ing, it uses its head and strong muscular neck a good deal, tossing the earth upwards and to the side, in a way that recalls the old name "moudie-warp" or "mould-tosser." Thirdly, it can run about on the surface at the rate of about $2\frac{1}{2}$ miles an hour, and the pairing takes place above ground. It must also be able to trot along in those underground runs which have some per-manence, e.g. the "bolt-run" from the headquarters. As to this so-called "fortress," it consists of a roughly spherical nest about the size of one's head, filled with leaves and grass. Above and around this resting-place there is a mound made of the earth which has been dug out, and traversing this there are tunnels or galleries which were made in transporting the excavated earth and may connect with the bolt-run or other radiating paths. No two "fortresses" show the same plan of galleries; their symmetry and significance have been exaggerated; they are simply the necessary outcome of making a comfortable resting-place and piling up the excavated material. According to some naturalists, an elaborate "fortress" is made by the males only. The sexes live apart, and the well-lined nest made by the female in May is usually under an inconspicuous hillock. The young ones, usually four or five in number, are pink and naked to start with, and very helpless. But the development is unusually rapid, the infantile period being telescoped down, and the offspring are able in five weeks to follow their mother and begin mining. The full-grown males are very combative; indeed, there is a good deal of suppressed fury in any mole. Everything they do is done with vigour and zest—moving, feeding, fighting, everything. A mole has been known to displace a nine-pound brick on a smooth surface, which for an animal weighing three ounces is equivalent to a man of twelve stone moving an object weighing 3 tons 12 cwt. (Frances Pitt, *Wild Creatures of Hedgerow and Garden,* 1920).

The Mole's vigour must be correlated with its extraordinarily good digestion. A mole can easily dispose of its own weight in earthworms in a day, and adults require food every three or four hours. A mole that was fed with forty earthworms late in the afternoon was found dead next morning with an empty stomach!

§ 6

Arboreal Mammals

Whether the earliest mammals were arboreal or not, it is a mode of life which many have adopted, and it has obvious advantages of increasing the freedom of movement, of securing a relatively safe retreat, and of making a nest a possibility. In many cases, as in a wild cat, the sharp claws are well suited for holding on to the branches. The Squirrel runs up the trunk, gripping with its claws, but looking as if it did not need to hold on; and its bushy tail is of use as a rudder when it takes an adventurous leap from tree to tree. In some cases, however, there are specially attaching structures; thus the extraordinary lemur called the Tarsius Spectre has disc-like suckers on its fingers and toes. Sometimes there is a splitting of the hand and foot which gives the limb a secure grip of the branch, and the same result may be reached by having an opposable first digit, like our own thumb. The Tree-Sloths show yet another method, for their claws are greatly elongated into hooks, and by means of these they move cautiously along, back downwards, hanging to the underside of the branches. It is interesting to notice how many features of these strange creatures have been altered in relation to their upside-down mode of progression. Thus they can bend their head round so as to look downwards over their shoulder; the neck is very mobile, and in some species has nine instead of the usual seven vertebræ; the shaggy hair hangs down in a unique way, and its suggestion of a mass of fibrous plants may be enhanced by the presence of a green Alga. One of the most effective adaptations to arboreal life is the most familiar—namely, the prehensile tail of many monkeys. In the Spider Monkey (Ateles) the tail is used not only to support the whole body, but actually as a "fifth hand" for grasping the food. Again, we get an impression of the plasticity of animal structure—the same part being turned and twisted to so many different uses.

The Squirrel

It may be doubted if there is any climbing mammal with more all-round attractiveness than the Common Squirrel. It is small without being pigmyish; the bushy tail balances the body; the rich brown-red upper colouring is very pleasing; the ear-tufts present during the colder half of the year make the creature look even more alert than it is; its movements take one's breath away.

Its table manners are perfect, for it sits upright, holding its food daintily in its hands; it neatly unshells the kernel of the nut; it even removes the thin outer pellicle before it begins to munch. Everyone knows how the squirrel passes from tree to tree, but it may also press its body against the stem and remain perfectly still. When it sleeps it uses its tail as a blanket. The security of its life probably adds to the gaiety of its disposition, for it is one of the playing animals, enjoying what looks like "tig" among the branches. Squirrels usually pair early in spring.

Two or three blind and naked young ones are born in a large nest of moss and leaves and twigs, which the monogamous parents build among the branches. There is strong maternal care and courage, and when danger presses the mother may carry one baby after another in her mouth to some place of safety. There is considerable instruction in athletics and woodcraft.

When winter comes the Squirrel does not hibernate, though on a very cold morning it may sleep late within the hollow tree. It still finds seeds and shoots to eat, and when these are scanty it searches about for the caches of nuts it made in September and October—and forgot all about! Too much has been made of the Squirrel's thrift.

§ 7

The Aerial Mammals

Although the scanty fossil remains of Bats have revealed nothing as to their ancestry, it seems safe to say that they evolved from an Insectivore stock. Specialised as they are for flight, they show numerous affinities with tree-shrews and the like. The vacillating rapid flight is familiar, and in some bats the power of flight is enough to enable them to migrate as birds do.

In relation to the bat's twilight habits, the sense of touch is highly developed on the wing, and about the nose and ears, so that obstacles like branches are avoided. Perhaps there is a pressure-sense distinct from touch, for bats often swerve to one side before they are near the obstacle. It has been suggested that the bat's cry of short-length sound-waves has a sort of echo from surfaces, and that this warns the bat from collision. There is, usually, only one young one at a time, an important restriction for a flying mother that has to carry the offspring about with her after birth as well as before. The back teeth of small bats bear sharp cusps, well suited for crunching insects, and a crowning adaptation may be found in the winter sleep of the bats of northern countries.

Vampires

The large bats, sometimes called "flying foxes," ranging from Madagascar to Queensland, are all fruit-eaters. The small bats are typically insect-eaters, but some are carnivorous, a few take fruit, and a few are blood-suckers. In the Vampire (*Desmodus fusus*), which feeds on blood, the gullet is so narrow that nothing but fluid could pass down. In his *Edge of the Jungle* (1921) Mr. William Beebe gives a graphic description of the vampires of British Guiana. They entered the bungalow at night and flew about, fanning the faces of the inmates, but for a time never touching. Eventually one would settle down on an exposed foot or arm, and creep on it, pushing with the feet and pulling with the thumbs, after the usual bat fashion, but so gently that the only sensation was a slight tickling and tingling. All this was preparatory to a small bite which would not awaken a sleeper.

British bats are all insectivorous. They congregate in considerable numbers in trees, caves, roofs, and holes in towers; but the sexes usually live apart. While typically nocturnal, they are occasionally seen in daylight; and, similarly, while they typically hibernate in winter, they are often seen if there is a spell of mild weather at that season.

§ 8

Mammals of Deserts and Steppes

The essential quality of dwellers in the desert is a capacity for rapid movements—to find herbage in a new area, to get out of a dry and parched land, and to flee from enemies where there is no possibility of concealment. Thus it is profitable to have long legs, a strong heart, good wind, and keen senses. The fleet Antelope may serve as a type, and there is a touch of perfection in the elusive Jerboa. Its long jumps must be disconcerting to an enemy, and the tuft of strong hair on the foot keeps this attractive biped from sinking into the loose sand, when it alights from its flying jump. Another feature, well illustrated by the

Photo: British Museum (Natural History).

"FORTRESS" OF THE MOLE (in cross section)

The large mounds of earth noticeable where moles are abundant are the "fortresses," while the smaller mounds are made by the earth thrown up during the construction of the "runs." The sexes build separate fortresses, and that of the female—an example of which is shown here in section—is the larger, being used as a nest for the young.

Photo: W. S. Berridge.

TWO-TOED SLOTH OR UNAU

This old-fashioned type (*Cholœpus didactylus*) lives in the forests of South America, e.g. in Nicaragua. It is highly specialised for arboreal life, moving slowly about back-downwards along the under side of the branches, holding on with the recurved claws on the two fingers and three toes. On the ground it moves awkwardly. It feeds on leaves, and has a stomach with several chambers. The hair is coarse and shaggy and affords a basis for the growth of a unicellular green alga. The teeth are simple pegs without enamel, and seem to be confined to one set. The two-toed Sloth has usually six neck vertebræ and the three-toed Sloth has usually nine, thus illustrating divergence from the normal mammalian number seven.

AMERICAN GREY SQUIRREL (*Sciurus carolinensis*)
COMING DOWN A TREE

An attractive and beautiful native of North America with habits similar to those of the Red Squirrel. A large nest is built on or in the tree, and there are usually two litters in the year. The creatures show great enthusiasm in hiding stores, including single nuts, in the ground. When pursued they press themselves flat and quiet on a branch, or take daring leaps from tree to tree. Many small colonies have established themselves, sometimes from Zoological Gardens, in Britain; and the diffusion on the shores of Loch Long shows the danger of introduction. The animals are very delightful in the London parks, but they may do enormous damage in woods and forests. The pet of confinement is apt to be the pest of the open.

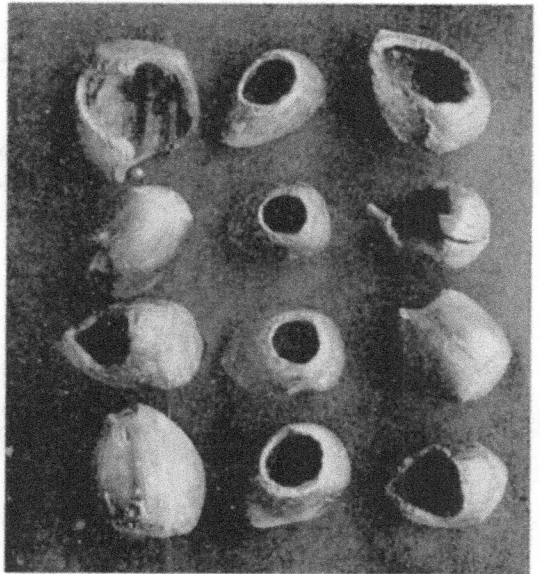

NUTS GNAWED BY SQUIRRELS

The holes in the empty shells show the neat work of the chisel-edged incisor teeth, and also, in most cases, the marked economy, for the aperture is not made larger than is necessary to let the kernel out.

Gazelle, is the "spareness" of build; the limbs are "all muscle"—
"muscle as hard as steel." There is, however, great elasticity
in the skeleton of the fore-limbs and in the connection of the
shoulder-blade to the backbone. It is easy to interpret the reduction in the number of digits as a lessening of friction, and the
same might be said in regard to the transformation of claws into
hoofs, but some of the peculiarities of desert animals are not so
easily explained. Are the markedly swollen nostrils of gazelles
and their relatives adapted to facilitate respiration in their racing,
or have they to do with filtering the air from the driven sand?
Opinions seem to be very discrepant in regard to the protective
value of the coloration of desert animals. A sandy-brown shade
is certainly very common, and apparent exceptions, such as
zebras, may admit of ready explanation. In the open the zebra
can look after itself and show quick heels; in the oasis it may be
that the striping is very inconspicuous. It is said that the huge
giraffes are very inconspicuous in a grove of acacia-trees.

The two-humped Bactrian Camel and the one-humped
Arabian Dromedary show various fitnesses for sandy deserts.
Thus the two toes have short nails instead of hoofs, and are
almost embedded in a strongly developed expansible sole-pad
with an elastic cushion between it and the bones. The result is
a surface which expands under pressure, and is well suited for
moving over the loose sand. In the closely related Llamas from
the Andes each toe has its own sole-pad, which is adapted for the
mountain paths. Many desert animals can go for a long time
without food or drink, and this is especially true of dromedaries.
In the paunch of these animals, and in camels, there are numerous
side-pockets with narrow openings which can be closed by circular
muscles, and these become filled with fluid. But we must not
make too much of this, for the water-pockets are also seen in the
Llama. Indeed, there are traces of them in the American
Peccary, which is related to the family of pigs. What has happened in the case of the Camel and Dromedary is probably that

special and adaptive use has been made of what was already present apart from desert conditions altogether. More unique is the development of a hump or of two humps, consisting chiefly of fat. When the animal obtains for a time a considerable quantity of moist herbage, the hump stands up tensely; when supplies are scanty the hump is reduced in size and becomes flabby. Another adaptation may be found in the Camel's power of completely closing its nostrils during a sand-storm.

Mountain Mammals

Really great mountains often show three zones—of forest, of steppe-land with scanty vegetation, and of barren grounds or tundra in the higher altitudes. Thus we find, among mountain mammals, forest forms like Bears and some Monkeys, steppe forms like Chamois and Yak, and tundra forms like Marmots and Snow-Voles. Many of the mountain mammals are of very hardy constitution, with thick fur, with great climbing powers, and with a capacity for enduring severe conditions and a starvation diet. Many are refugees from the low grounds, and some, like the mountain beaver, are very old-fashioned, primitive types.

The Mountain Hare

The Variable or Mountain Hare is a first cousin of the Common Hare, and is nowadays a distinctively northern mammal. When the ice-sheet was thick over the mountains of Central Europe the Variable Hare lived in the low grounds. When the climate became milder it had to retreat—either further north or up the mountains. It became extinct in England, but has been re-introduced with success. Compared with the Common Hare, it is smaller as a whole, and in its head, ears, hind-legs, and tail; its flesh is whiter; it is a less dainty feeder. It does not seem to have any particular home or "form," but shifts about restlessly from one hiding-place to another. When the snow is deep it is forced to descend to lower levels. In Scotland it usually turns to white in winter, all but the black tips of its ears; in Ireland there is not usually any seasonal change of colour.

The Story of the Snow-Mouse

One of the most definitely mountain-haunting mammals is the Snow-Mouse, or, accurately, Snow-Vole (*Microtus nivalis*) of the High Alps. It is a little creature about five inches long in body and two more in tail, usually rusty-grey or whitish-grey in colour. Perhaps it has the honour of living a harder life than any other mammal, for it is rare below 4,000 feet, and it ascends from the snow-line to the tops of the mountains. It does not migrate in winter; it does not hibernate; and it does not turn white. In fact, its only adaptation to its snowy retreats is that in the summer it gathers to its nest among the loose rocks a store of chopped grass and gentian roots. In winter it makes tortuous burrows beneath the snow, mining its way from one Alpine plant to another. It has the reward of freedom from enemies, for even birds of prey are scarce at these heights. The explanation of the habitat is interesting. The snow-mouse used to be one of the "tundra" animals, like the Reindeer and Arctic Fox, that frequented the low grounds of Central Europe when the uplands were covered by a great ice-sheet. As the climate became milder and the ice-sheet melted, some of the "tundra" animals, like the Reindeer and Arctic Fox, retreated northwards, but the Snow-Mouse went up the mountains, higher and higher. Thus we also understand why they have to-day a scattered distribution, separated by extensive mountainous tracts where none occur. This corresponds to some extent to separate migrations from the low grounds; it also has to do with the available vegetation, for the hardy Snow-Mouse must eat something.

§ 9

Mammals show a thousand and one adaptations connected with procuring and utilising their food, and we cannot give more than a few illustrative examples.

Food-getting among Mammals

The Great Ant-eater (Myrmecophaga) of South America comes out at night and with its exceedingly powerful claws breaks into the earthen hills of the termites. Then out and in whips the thread-like sticky tongue, drawing hundreds of insects in a short time into the absolutely toothless mouth. The same kind of tongue is seen in other ant-eaters, such as the Aard-vark of South Africa, and in the oviparous Echidna, which is also absolutely toothless.

The whalebone whale, of whatever kind, swims open-mouthed through the surface waters, engulfing myriads of small sea-snails and the like in the huge gaping cavern. The small animals are caught on the frayed edges of the baleen plates, exaggerated horny ridges of the palate, which hang downwards from the roof of the mouth. Every now and then the whale raises its tongue and brushes a multitude of the entangled creatures towards the back of its mouth, where they are gripped by the pharynx and swallowed. The water streams out at the sides of the mouth through the sieve of whalebone, but some of it would be apt to "go the wrong way" were it not that the whale shunts its glottis (the opening to the windpipe) forward to embrace the posterior end of the nasal passage. What a contrast is such a mouth to that of a toothed whale, like the Sperm-Whale and the Dolphin, with teeth well suited for seizing cuttlefishes and fishes! Yet it is interesting to notice that the whalebone whale has before birth two sets of teeth, which never cut the gum!

The adaptations of the teeth of mammals to different kinds of food-getting are many; but from a few we may learn all. In the gnawing mammals or rodents, such as rats, beavers, porcu-pines, and squirrels, the enamel is either confined to the front of the incisors, or it is much more strongly developed in front than it is behind. Thus the posterior part of the tooth wears away faster than the anterior part, so that a chisel edge is automatically formed. The lower incisors strike in behind the upper ones, and this keeps the enamel edge sharp. Moreover, these teeth are "rootless," and go on persistently growing as they are worn away. In the gap behind the incisors, where canines should be, an infolding of the skin into the mouth cavity separates a front portion from a back portion. Thus material which is being gnawed, but not intended to be swallowed, may be kept from going beyond the front region of the mouth. Some of the rodents, like the Gopher, store what they gnaw in capacious cheek-pouches, and grind this with their back teeth when they get into a place of safety.

No one can look at an elephant using its trunk without recognising a new idea—the employment of the nose (and a prolongation of the upper lip as well) as a food-getting organ. This is Nature's way, making an apparently new thing out of something very old; and it is evident from the remains of extinct elephants that the trunk or proboscis had a gradual evolution, proceeding in correlation with that of the huge tusks which prevent the mouth getting close to things in the usual way. The efficiency of the trunk is greatly increased by a very mobile, finger-like process at the tip, which enables the elephant to handle little things as well as to lift great logs.

The trunk of the Elephant is a masterpiece, and the initial stages may be discerned, not only in the evolutionary history, but in the short proboscis of the Tapir, and even in the sensitive snout of the Pig, which is used for routing in the earth in search of food. There is a special snout-bone (pre-nasal) in pig and mole; but the risk of hasty interpretation in terms of fitness may be illustrated by the fact that the same bone occurs in the Tapir, which does not rout in the earth, and also in Tree-Sloths! The bone in question is probably a primitive feature, for the Tapir, for instance, is a very archaic mammal. In some cases, like the Elephant Shrew, the proboscis is a puzzle: we do not know its use.

The Elephant

The Elephant type, now represented by two species, the African and the Indian, exhibits many zoological peculiarities besides the familiar trunk and tusks. Thus the limbs are quite unique among living mammals in their straightness; they form vertical pillars adapted to support the huge weight of the body. But there is even greater interest in the ways of the creature. According to Sir Samuel Baker (*Wild Beasts and their Ways*, 1890), the African Elephant can charge for a short distance at the rate of fifteen miles an hour, and keep up the rate of ten miles an hour for a long run. The tusks which form the weapons of the males in their furious combats are used by both sexes in everyday life for digging up roots for food. It is said that an elephant does not reach proper maturity till it is forty years old, and that it may live far over a century. It is one of the slowest of breeders and carries its young for twenty-two months before birth. Yet we recall Darwin's calculation that after a period of 750 years there would be nearly nineteen million elephants alive, descended from a single pair. The cerebral hemispheres of the big brain are richly convoluted, and the creature is so intelligent that "elephant stories" are proverbial. Of its memory, of its capacity for learning both in peace and war, and of its practical judgment, there is no doubt.

Chewing the Cud

Some of the hoofed animals, such as cattle, sheep, and deer, illustrate an interesting peculiarity called chewing the cud, or rumination. These animals feed, as everyone knows, on grass and herbage, and it is often important for them to eat as much as they can in a short time. A choice patch must be utilised to the full, and there is always the danger of an attack from carnivores. So the ancestors of our sheep and cattle got into the habit of gorging themselves with hastily swallowed grass, and then of retiring to the place of safety—often with their backs against a rock so that

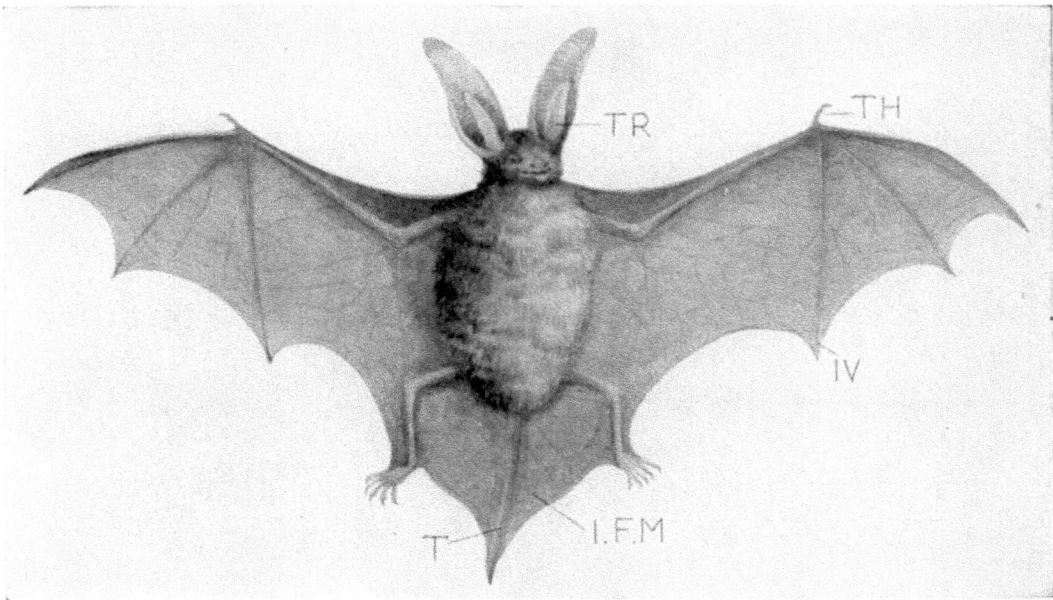

THE LONG-EARED BAT

Long-eared Bat (*Plecotus auritus*), a common British bat, which does good work in destroying injurious insects. It is a playful creature, not difficult to tame. The ear-trumpet is larger in proportion than in any other animal, and may be 1 ½ inches long. The body, not counting the tail, is about 2 inches in length. In the ear, as in many other bats, there is a very strong development of the forward flap or tragus (TR), which is represented by a small anterior lobe on our ear-trumpet, guarding the entrance. The projecting clawed thumb (TH) is clearly shown, and the fourth digit (IV), corresponding to our little finger. Between the hind-legs, supported by the tail (T), there is an inter-femoral membrane (I.F.M.), a basket of skin against which the bat presses its insect booty when it is killing them during its flight.

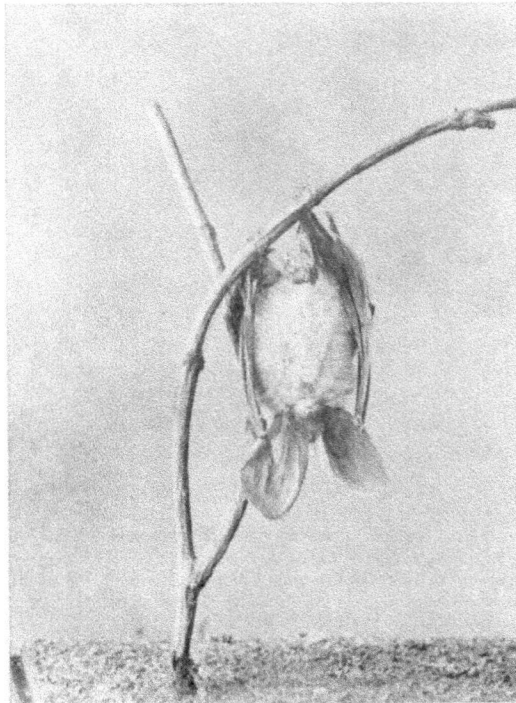

Photo: Aberdeen University Museum.

LONG-EARED BAT IN RESTING ATTITUDE

In the Long-eared Bat (*Plecotus auritus*) the ears are nearly as long as the body! When the animal is resting squat, it folds its long pinnæ backwards along its body, and that leaves projecting forward from each ear a strongly developed earlet, or tragus. When the bat hangs itself up to rest (as above) the long ears hang down.

THE JERBOA. (*From a Specimen.*)

A Jerboa (Dipus), a biped mammal, adapted for life on the deserts and steppes. The ankle joint is very high off the ground; the foot has a tuft of hair which prevents it sinking into the soft sand; the tail is useful in balancing; the fore-limbs are bent close up to the neck. The length of leap is extraordinary, and the creature vanishes almost instantaneously. Jerboas are also able to burrow. An astonishing feature is the coalescence of the three instep or metatarsal bones into one, presenting a strong resemblance to the tarso-metatarsus of birds. More remarkable still is the "soldering" together of the neck vertebræ. Jerboas of this genus are confined to the Old World.

they could not be surprised from behind. There, at leisure, they re-chewed their hasty meal.

The so-called stomach of a typical ruminant, such as sheep or cow, consists of four chambers. The first is the capacious paunch or rumen, the internal surface of which is thickly beset with tag-like processes, suggesting velvet pile. It is here that the grass is stored; it is acted upon by the salivary juice which has followed it down, and there is also some bacterial fermentation. The second chamber, the honeycomb bag or reticulum, is marked by a hexagonal pattern, and it rarely contains more than sappy fluid. The third chamber, the manyplies or psalterium, has numerous plaits filling up its cavity, so that the food has to pass through a kind of filter. The fourth chamber, the reed or abomasum, is the seat of gastric digestion. In fact, it is the true stomach, for the preceding three chambers turn out to be elaborations of the lower end of the gullet or œsophagus. This is known by the minute structure of their walls, for there is no confusing the non-glandular gullet region with the very glandular stomach region.

What happens in rumination? The cow, lying slightly on one side, returns boluses of food from the paunch to the mouth, where they are very thoroughly masticated and moistened with saliva. If we watch a cow we can see these boluses or rounded masses of vegetable matter travelling up the gullet with considerable rapidity. After the thorough chewing, the food is re-swallowed and passes down for the second time; the muscles of the gullet working in a manner the exact opposite of that exhibited when the boluses pass up. On the second descent the food skips the paunch and the honeycomb bag, there being automatic arrangements for preventing entrance, and travels along a groove into the manyplies. Filtering through this third chamber, it reaches the true stomach and is subjected to gastric digestion.

Overloading a stomach sometimes leads to vomiting—an

automatic means of getting relief—and although the paunch is
not stomach it is difficult to refrain from suggesting that the first
part of the cud-chewing process may be a sort of normalised
vomiting.   Nowadays the whole series of steps is reflex or automatic, but it is interesting to notice that if the cow is disturbed in
the middle of its cud-chewing it is not a little put about, and is
often unable to resume the process for a considerable time.   Such
disturbance is of course injurious to the animal's health.

§ 10

Weapons of Mammals

Many mammals use their teeth, especially their canines, as
weapons.   The Walrus strikes downwards with tremendous force,
the Wild Boar lunges upwards with the canines of both jaws
pointing up.   In the Asiatic Babirusa the upper canines, though
pointing up, are curved so far backwards that they form more
of a shield than a weapon.   In male Musk-Deer, and in the likewise hornless but quite unrelated Deerlets, the canines are
strongly developed and are used in fighting; in Elephants the
great tusks are front teeth or incisors.   The use of the six-foot
long left canine of the male Narwhal remains obscure.

Besides their teeth mammals may use as weapons their claws
and their hoofs, and various kinds of horns.   The Rhinoceros has
a horn that belongs wholly to the skin—like a huge wart that has
become very hard.   The horns of cattle, sheep, and deer have a
core of bone (growing from the forehead or frontal) covered by
an integumentary hollow sheath of horn.   In the Giraffe and the
Okapi the sheath over the bony outgrowth does not turn into
horn.

The Story of Antlers

Antlers deserve a place by themselves.   They are restricted
to stags with the single exception of the Reindeer, where they
occur in both sexes.   They are not seen in the buck's first year,

when there is only a small, permanent, skin-covered, bony out-growth on the forehead, called the pedicle, which grows in girth in subsequent years. In the second year there is an extra-ordinarily rapid multiplication of bone-forming cells on the top of the pedicle, and a short unbranched antler is formed, which carries upwards the hot skin or "velvet." The blood-vessels in the velvet supply the food which admits of the rapid growth of the skin, and they also keep the growing antler tissue suitably warm. The materials for the growth of the antler itself are brought by internal blood-vessels from the pedicle or stalk. Branches from the fifth brain-nerve run up the velvet and make it exquisitely sensitive—an adaptation that saves the stag from knocking the still soft antlers against hard objects.

In ordinary deer the antlers are as transient as the leaves of the forest. They drop off and there is a new growth next year. The second antler has a stem and one branch or tine, and a new tine is added each successive year until the stag reaches maturity, after which the antler growth becomes irregular.

The shedding of the antlers is an extraordinary process. It is prepared for from the start by automatic arrangements which cut off the supply of blood from the velvet, obliterate the internal blood-vessels, and form at the base a soft tissue which loosens the organic connections between the dead antler and the living ped-icle. The dying away of the base of the antler would be called disease in other animals; it has become mysteriously regularised in stags. The whole process is extraordinary; the growth of a fine "head," perhaps 70 lbs. in weight, takes place in three months —an expensive utilisation of material called into activity by chemical messengers (hormones) from the reproductive organs. The splendid result is hardly finished before operations begin for its being shed! And after all, the antlers do not seem to be of much practical importance; they are exuberant outcrops of the male's virile constitution. Perhaps they have their counterpart in the male narwhal's spear.

The Red Deer

Britain has lost the Reindeer and the Giant Deer, a fine creature of the ancient forests, but it still has the Red Deer (*Cervus elaphus*), which is genuinely wild in some parts of the country. It stands about four feet high at the withers, and the veteran stag has truly magnificent antlers, which are called "royal" when they have over twelve "points" or branches. The stags are very combative at the breeding season (September and October) and may be dangerous to man. They are greatly excited and roar loudly, challenging other males. In their ferocious combats they push with the antlers as a whole, or they stab at the heart and belly with the lowest branch or "brown-tine," which points forwards and upwards. A good deal of use is also made of the hoofs, especially those of the fore-feet. Each stag tries to attach to himself as many females as he can. The fawn is born in May or June, spotted as in most deer; it is carefully guarded by the mother, who teaches it to conceal itself when it hears the danger-signal—a tap with the fore-foot. In the summer months the hinds and fawns usually live apart from the stags, and often at a lower level. Although we associate the Red Deer with the Highland hills, to which they are well adapted in their strength and swiftness of limb, in their close-set coat, and in their wonderfully keen senses of smell, sight, and hearing, they were originally forest mammals rather than mountain mammals. They feed mainly on soft grass and heather shoots, but they have interesting vagaries of appetite such as gnawing at their cast-off antlers. Like the Reindeer of the Far North, they sometimes travel a long distance to get an early morning lick at the rocks on the seashore.

Protective Adaptations

Some of the archaic mammals show a remarkable development of armour. The Armadillos are unique in having a bony skin-skeleton which is almost invulnerable, especially when the animal rolls itself up, thanks to the flexible rings in the armour, into an unopenable ball. Very striking is the tiny Pitschiago from barren grounds in South America. It has a bony carapace above, and on its under parts very beautiful snow-white hair; it has enormous nails on its fingers by which it is able to burrow very rapidly; and its hind-parts have a special very decorative shield. Hardly less striking are the Pangolins (Manis) with the body covered with very hard overlapping scales of horn, suggestive of a reptile rather than of a mammal. There is an Indian Ocean porpoise which has calcified scales all along its back, and, as these are larger before birth than after, it seems safe to interpret them as legacies from a very distant scaly ancestry. It seems that our Common Porpoise has sometimes very hard tubercles in its skin, and perhaps this also illustrates the hand of the past living on in the present.

But there are other kinds of armour besides scales. The Porcupine has its long spines, the Hedgehog its short ones, and the Spiny Ant-eater is intermediate between the two. Even in the hide alone there may be considerable strength of armour—as in Rhinoceros, Hippopotamus, and Elephant. In many cases no armour is required, for the creature is endowed with relative invisibility, as we have seen in a previous chapter.

§ 11

Nocturnal Mammals

Many animals of long pedigree have adopted a nocturnal mode of life, which gives them additional safety in circumstances more difficult than those to which they were primarily adapted. Thus the Otter and the Badger owe their survival partly to their nocturnal habits, but it cannot be said that they are in any very marked way adapted to walking in darkness.

The Story of the Badger

The Badger (*Meles taxus*) has still a firm footing in various parts of Britain, such as Devon and the New Forest. It is a thick-set, round-backed, rather bear-like carnivore, somewhat over two feet in length, with an additional seven inches of tail. It has a long muzzle, well suited for its restlessly inquisitive poking into holes and corners; the short rounded ears are not in the way in the brushwood; there are bright bluish-black eyes; there is below the tail an odoriferous gland with a disagreeable smell. The Badger stands alone among British mammals in having the under parts darker than the upper, for the under surface is black while the upper surface is tawny, overlain with grey, darkening here and there. The head is practically white, divided by a broad black band beginning between the nose and the eye and extending back to the ear. In short, the colouring is rather conspicuous, recalling the American Skunk. But the Badger is elusive, and though it has few enemies it will work its way in the dusk down a dry ditch or along the side of a hedgerow rather than across the open. The heavy body does not seem to be lifted much off the ground, the snout is often held very low, the soles of the feet are entirely on the ground in true plantigrade fashion. Yet the badger's movements have an easy swing, and the creature does not know what it is to be tired.

When we ask how the Badger manages to survive in a much cultivated and far from friendly country, part of the answer is in the words nocturnal and self-effacing, and, possibly, evil-smelling. We must add, however, that the Badger has strong positive qualities. It is very muscular; it has a strong heart and a good wind; the grip of the lower jaw is unsurpassed in tenacity; the thick coat helps the badger to withstand the cold of winter; it stores a good deal of fat; it is endowed with keen senses, shrewd intelligence, and a capacity for taking things easily without fuss or worry. And yet this is not all. It has an extraordinary catholicity of appetite, which always makes for survival. If one kind of food fails, it can fall back on something else—roots and fruits, nuts and truffles, worms and grubs, frogs and snakes, eggs and young rabbits, the grubs from the wasps' nest (for the badger is

THE MIDDAY HALT

Camels are represented to-day by the two-humped Bactrian Camel (*Camelus bactrianus*) and the one-humped Dromedary (*C. dromedarius*), and by the geographically far distant Llamas (*Lama*) of South America. Neither of the Old World forms is now known to occur in a thoroughly wild state. Herds that have gone wild or become "feral" are well known. The spreading out of the third and fourth digits is adapted for treading on soft sand. The paunch has got "water-cells" and smooth walls. A quite unique feature is that the red blood corpuscles are elliptical in contour, instead of circular as in all other mammals.

Photo: James's Press Agency.

GREAT ANT-EATER

The Great Ant-eater (Myrmecophaga) is one of the South American terrestrial Edentates. There is no hint of teeth, but insects are caught by the rapid protrusion and retraction of a worm-like sticky tongue. The powerful claws are used in tearing up the ground or breaking into ant-hills; they are formidable weapons besides. There is shaggy greyish-black hair over the body, a broad white stripe on the shoulders, and a big bushy tail, only partly shown in the photograph. The length from the tip of the greatly elongated snout to the tip of the tail may be over 7 feet.

Photos: by Courtesy of Charles Hose.

THE TAPIR OF SUMATRA (*Tapir indicus*)

It will be noticed that the young one (upper photograph) is striped and
spotted. These stripes disappear during the first year, giving place to a
well-defined black and white pellage when fully adult (lower photograph).
The young one, with its yellowish spots and stripes, is "like a patch of ground
flecked with sunlight"; the adult with its two colours is like a grey boulder.
Tapirs form a small family of hoofed mammals (Ungulates) related to rhino-
ceroses and horses. Their modern geographical distribution indicates great
restriction compared with that in bygone ages, for some of the species occur in
the Far East, the others in South and Central America.

impervious to stings), and the honey from the humble-bees' store. Another factor is its burrowing habit, for its "earth" or "set" goes far in and may have several entrances. It is made comfortable with bracken and herbage, and is kept fairly clean. Moreover, one must attach survival value to the education which the mother badger gives to her silvery-grey cubs. There are usually just two or three of them, born in spring. When they have got their sight, some ten days after birth, and had their usual gastric education on milk, they are taken outside the warren and well groomed. Then comes schooling, and the mother is a stern disciplinarian. She punishes the inattentive and foolhardy, and gradually instructs them in the way in which they should go.

The Hedgehog

The Hedgehog is an old-fashioned insectivore that holds its own well from Britain to the Ural Mountains. It does so in virtue not of brains or of weapons, but because of other fitnesses. Many of the hairs have been transformed into sharp spines, which are erected by the smooth muscles at their base when the animal is touched. They also serve to break the force of a fall when the Hedgehog, a good climber, tumbles from a wall or a tree. A very strong dome of muscles beneath the skin (see cut facing p. 491) rolls the animal up into an unopenable ball. The senses are acute; the prolonged snout is well suited for probing into holes; there is a wide range of appetite—earthworms, grubs, slugs, and small snails; and the mountain-top-like cusps on the back teeth are well suited for crunching these. The constitution is very tough, and if the Adder—an inveterate enemy of the Hedgehog— gets a bite in, the venom has no effect. Experiments with poisons and with such germs as that of diphtheria have proved the refractoriness of this common creature. Although it has few enemies, it adds to its safety by resting during the day in a well-hidden recess, and hunting by night. There are often two litters (usually of three or four) in the year, and the young one is a curious flat and feeble creature, with soft white spines pointing backwards, and a pale blue-grey skin. It is not for some time able to roll itself up, yet it develops quickly, and is able to follow the mother in a month or two.

§ 12

Hibernation

Many creatures, such as reptiles, amphibians, snails, and insects, pass into a lethargic state when winter sets in, and lie low until the spring. But it is only in mammals that we find true hibernation, a very peculiar physiological condition, which is not sleep, nor *necessarily* connected with winter. It is exhibited by Hedgehog and Hamster, Dormice and Bats, Marmot and Souslik, the Spiny Ant-eater of Australia and the Jerboa of the Kirghiz steppes.

To understand the hibernation or so-called winter sleep of these mammals, it is necessary to recall the main facts in regard to animal heat. Inside the body heat is produced by various chemical processes, but mainly by the muscles; it is of great importance in facilitating the operations of the living laboratory. But the heat tends to be lost by radiation into the outer world through the skin, and in the hot breath and in sweating. The non-conducting fur in ordinary mammals and the blubber of whales lessen the loss from the skin, as do the feathers of birds. But there is in birds and mammals a self-regulating system, which keeps the temperature approximately constant, day and night, year in and year out; and this is what is meant by warm-bloodedness. The regulating centre is in the brain, whence orders issue to the muscles, blood-vessels, and skin. If too much heat is being produced or lost, an adjustment is effected. But all mammals are not perfect as regards this heat-regulating arrangement, and it is among these that hibernation occurs. A good example may be found in the Spiny Ant-eater (Echidna), whose temperature may vary ten degrees Centigrade according to that of the outside

From a photograph by J. E. Russell, Dingwall.

STAGS FIGHTING WITH THEIR FORE-FEET

The Red Deer stags grow and shed their antlers annually. During the period of growth the horny substance is traversed by nerves and blood-vessels and is covered by a soft, sensitive sheath of "velvet." When the antlers are fully developed they become hard, non-living structures and the "velvet" dies and is rubbed off. A few months later the whole antler is shed and the process begins anew. The fiercest combats are in the mating season, and then the antlers are used, but when the new horn is still "green"—and again when the old antlers are being shed—fighting is with the fore-feet in the manner shown. (See *Natural History: I!. Mammals*, page 331.)

world, whereas our temperature varies only by a fraction of a degree as long as we are in good health. Now the Spiny Anteater is a hibernator, and this is the clue we need: winter-sleeping mammals are imperfectly warm-blooded. When the cold weather sets in, it becomes difficult for them to adjust the debtor and creditor account as regards heat; they cannot produce enough to make up for their loss, and they give up the attempt. They sink back into a state of comparative coldness and cold-bloodedness; they relapse into the ancestral reptilian condition.

But if the imperfectly warm-blooded mammals which we have mentioned were to fall asleep in the open, their body-temperature would go down and down, and they would die. What they must do is creep into some sheltered nook or comfortably blanketed hole where the temperature soon becomes much higher than that of the world outside. To this temperature that of the sleeper's body approximates without there being any fatal results.

Along with the snuggling into a confined space must be taken the great reduction of internal activities, and here hibernation approaches the lethargy of frog and tortoise. Income is *nil*, so expenditure must be reduced to a minimum. The heart beats feebly, the breathing movements are scarcely perceptible, the excretion or filtering which is the work of the kidneys comes to a standstill. The hibernating body is like a fire well banked up in its own ashes, and in an animal like the Hedgehog we know that subtle changes come about in the recesses of the tissues.

The gist of the matter is to be found in the three facts: (1) a constitutional imperfection in the temperature-regulating arrangements; (2) a creeping into a confined space which gets warmed up a little; and (3) a great reduction of expenditure, for even the internal activities come almost to rest. But there are some contributory influences which must be recognised. After the hard work of summer, there is naturally some fatigue and a bodily bias towards rest. Moreover, summer has often been a time of plenty, and the body has accumulated stores of fat and other reserves, which may also incline the creature to somnolence. And once the quiescence has begun, it will tend to continue, for the closeness of the retreat must be soporific, and the cessation of the kidney functions will tend to keep the sleepers sleepy. Just as drowsiness sometimes sets in when man's kidneys are not working rightly, so in the hibernating mammal there may be a poisoning of the body with its own waste-products—a sort of "auto-intoxication."

Yet this is not all. We must not think of hibernation as an individual reaction merely; it expresses a racial rhythm. In the course of thousands of generations a certain periodicity has been established, like that of our sleepiness at night and wakefulness in the morning, and with the enregistered bodily rhythm there is associated an instinct which prompts the hibernator to seek out a comfortable corner when the weariness or sleepiness sets in. For ages, it must be remembered, our hedgehogs have not known any winter. They have slept through them all, just as the migratory birds have circumvented them all. It must be remembered, too, that the winter-sleep or hibernation of an animal like the Hedgehog cannot be distinguished from the summer-sleep or æstivation of the Tenrec of Madagascar.

Only a few mammals are hibernators, and some of these, like the Dormouse, are "light sleepers," while others, like the Hedgehog, are "deep sleepers." In all cases there is some imperfection in the warm-bloodedness, and what has been wrought out is what we might call a rather neat way of making a strength out of weakness. There is a relapse to a reptilian condition, but this handicap is counteracted. For it is not merely that the difficulties of the winter—scarcity, cold, and storms—are circumvented; the hibernation gives an opportunity for a long rest, which even the food-canal may be the better for. There may be an opportunity for processes of recuperation or rejuvenescence to stave off the processes of senescence or ageing. Why, then, are there not more hibernators? The answer must be that hibernation is the "answer-back" made by certain creatures with a constitutional peculiarity; other mammals meet the winter in other ways.

§ 13

Sex Dimorphism

The contrasts between lion and lioness, between stag and hind, are familiar. They illustrate what is technically called sex dimorphism, i.e., a marked structural difference between male and female. The contrasted characters are called secondary sex characters, to distinguish them from primary sex characters, which have to do more or less directly with the reproductive function itself. The males are sometimes equipped with decorations—the manes of lion and bison, the beards of certain goats, the crests along the back of some antelopes, and the dewlaps of bulls. Or they may have weapons which are either absent in the females or represented in less exuberant development. Thus antlers are restricted to the males except in the case of the reindeer; the horns of bull and ram may be much larger than those of cow and ewe; the male narwhal has a spear-like tusk which is not developed in the female. There may also be differences in colour and in odour.

Darwin suggested that when the males fought for the possession of the females, as stags and antelopes do, the males with better weapons would prevail. As they would therefore have most success in leaving progeny, their strong qualities would gradually become racial characters; the males with poor weapons would be sifted out. In regard to sex decorations he suggested that the females would be most interested in, and would give the preference in mating to, the more handsome males, and that the race would therefore evolve in the direction of increased decorativeness. This is, in brief, Darwin's theory of sex selection, which is discussed in the article, "How Darwinism Stands To-day." But one point must be noticed here. If the quality of having strong weapons or of having handsome decorations is hereditarily transmissible, why does it not appear in the female as well as in the male offspring?

How can it be entailed on the male offspring only? The answer must be that the quality is handed on to both sexes, but that it cannot find expression except in a male constitution. Similarly, the foundations of milk-glands are part of the inheritance of both sexes, but normally their development is restricted to the females. There are items in the inheritance of both sexes which are like seeds requiring particular kinds of soil if they are to develop. The male character of antlers or of shaggy mane requires a masculine constitution (including the presence or absence of certain hormones) if it is to develop. This leads to the view that the secondary sex characters are in their origin bound up with the primary differences of constitution implied in maleness (sperm-producing) and femaleness (egg-producing) respectively.

All theory apart, we return to the facts: (1) that the male mammal is often markedly different from his mate, (2) that there are often fierce combats between rival males, and (3) that in certain cases the females seem to show a certain preference, being apparently more excited by some males than by others. It is probably the total get-up that counts rather than any individual item such as an extra long beard. The combats of rival stags are sometimes furious, and the antlers are occasionally interlocked with fatal results to both combatants. A male antelope sometimes punishes an upstart youngster so severely that the blood flows from many wounds

§ 14

Family Life

Some mammals are monogamous, others polygamous, and others promiscuous. The monogamous forms include the Chimpanzee, the Tarsius Spectre, the Hedgehog, the Elephant Shrew (Rhynchocyon), the Pangolin, some antelopes and small deer, and the Bandicoot (*Perameles obesula*). The polygamous forms include most deer and antelopes, wild cattle and horses, sea-lions, and elephants. Not infrequently the males live by

Photo: W. P. Dando.

INDIAN ELEPHANT

There have been many different kinds of elephants in the course of the ages, but there are now only two, *Elephas indicus* (above) and *E. africanus*. The African Elephant is larger with much bigger ears, with a more sloping head without the two rounded bosses of the Indian, and the tip of the trunk has two triangular processes, one above and one below. The Indian species is the only one that is used as a beast of burden.

MAY 10.

JUNE 6.

JULY 18.

THE GROWTH OF ANTLERS IN THE FALLOW DEER

The young bucks show the first traces of antlers in their second year, but they do not get beyond mere "snags." During the next four years they become finer and finer. The antlers of the old bucks are usually shed about the beginning of May, and the new growth begins soon afterwards (see May 10). The antlers are rounded at the base, but become flattened or palmated higher up, as the third figure (July 18) well shows. Common report credits the Romans with introducing the Fallow Deer into Britain, but fossil remains have been found in at least one British cave. The general colour in summer is yellowish-fawn above with rows of large white spots; in winter the spots almost disappear.

SCOTTISH RED DEER (*Cervus elaphus*)

This head from Inverness-shire though not very large, is one of great beauty; in symmetry and proportions it is almost perfect. It shows seven points. The first tine develops in the third year, so a stag with seven points would be nine years old. The antlers are usually shed in March, and seem to be eaten by the deer themselves. The new ones are completed by July or August.

Photo: *W. S. Berridge.*

NINE-BANDED ARMADILLO OR PEBA

This strange archaic type (*Tatusia novemcincta*) occurs in arid regions in South America, and extends into Texas. Between the shoulder-shield and the hip-shield there are nine movable bands, but the creature does not roll itself up into a living ball. It is only in Armadillos that plates of bone occur in the mammalian skin; above the bone there are epidermal scales of horn. The teeth are numerous small, blunt pegs, without enamel. The Peba can run quickly and burrow quickly; it uses its claws as weapons. It hunts for insects at night, or at dawn and dusk; during the day it keeps in its burrow, which may descend for 6 feet into the dry soil. Very remarkable is the fact that this Armadillo normally produces *quadruplets*—four embryos from one egg-cell—and these, as might be expected, are always of the same sex, either all male or all female.

themselves except at the breeding season, as in the case of elephants and stags. An experienced old female leads the hinds and the young; a male in his prime leads the stags. Where the pair do not live together throughout the year, and where the care of the family devolves wholly on the mother, the terms monogamous and promiscuous have not much meaning. There is pairing rather than mating. A female mammal may pair with one male one year and with another next year, or with several in one year. But our knowledge of these matters is sadly lacking in precision. It has to be remembered that in most cases the pairing time is sharply punctuated and of short duration.

It is interesting to inquire into family life among apes. The Gibbons (Hylobates) of southeast Asia are the smallest of the anthropoid apes, rarely over three feet high. But they have disproportionately long arms, the hands touching the ground when the animal stands erect. They are fond of swinging like acrobats below the branches with their arms above their head. They can swing clear for 12–18 feet with the greatest ease, and pass from tree to tree unwearyingly. During the day they keep to the tree-tops, especially on the mountain-sides; towards evening they come cautiously "waddling" down in the open ground searching for fruit. Their voice is extraordinarily strong, especially in the males, and not unmusical. They are sociable and talkative. The Orang (*Simia satyrus*) of the forests of Borneo and Sumatra stands about four feet high and is very strongly built. It is highly intelligent, but somewhat sluggish in habit, climbing slowly, keeping to the trees except at night, when it sometimes searches on the ground for fallen fruit. It uses its arms as crutches or goes on all-fours. It makes for resting purposes a sort of platform nest of branches, but it moves on and makes another every second day or so. The male orang lives apart; but the mother keeps her family with her for some time. The Chimpanzee (*Anthropopithecus troglodytes*) of African equatorial forests may be five feet high, but it is not so bulky as the Orang, and it is as good a

climber as the Gibbon. It makes a temporary platform or
resting-place among the branches. In disposition, it is lively and
playful; it is easily tamed, and has a plastic intelligence. The
Gorilla, also restricted to Tropical Africa, may be a little over five
feet in height, and is of enormous strength in shoulders and arms.
It goes much more on the ground than any of the other anthro-
poids, and has a shuffling, rolling gait, using the hands a good deal,
and keeping the body semi-erect. It fights ferociously with hands
and teeth, and does not retreat from man. It is said to be gloomy;
it beats on its breast when enraged; it has never been tamed. A
single adult male usually leads a small company of females and
young ones.

Care of the Young

Some young mammals are born very helpless—blind, naked,
and with little power of movement. This implies some sort of se-
clusion or shelter, such as a burrow or a nest, as in Fox and
Squirrel respectively. In the case of the Rabbit there are both,
for the mother makes a bed of her own fur. During the very helpless infancy, the mother mammal is assiduous beyond telling. In
some cases, after a period of suckling, the mother brings animal
food to her young ones, and that food is not always dead. For the
education has to begin early. The play of the kitten (and even
of the cat) with the mouse is doubtless wrapped up with the
business of early education.

In some cases the young ones are carried about by the mother.
Reference has already been made to the marsupials, but there are
other instances. A mother hippopotamus is sometimes seen in
the Nile with a calf astride on her short neck: the young are
precocious, and the mothers very affectionate. Many monkeys
carry their babies about with them among the branches, and so
does the quaint Tarsius, which belongs to the order of Lemurs.
Among bats the young one is carried by the mother as she flies,
and the holding on is assisted by the front teeth which grip the

rough hairs. On a somewhat different line are the cases where the mother takes a young one in her mouth and transports it to a place of safety. This is familiar in the case of a cat and her kittens, but the squirrel may also shift her young when danger threatens.

In some cases the instruction given by the mother is an important factor in securing the survival of the young ones, and therefore of the race. Thus the Badger instructs its offspring in the art of being elusive and in the diverse ways of securing food. Even better known is the Otter's schooling, for the young are taught all the alphabet of country sounds, how to dive without splashing, how to lie hiding under the bank without betraying themselves, how to catch frogs and skin them, how to guddle for trout and eels, how to eat the eel from the tail and the trout from the head, how to deal with rabbit and moorhen, and how to find their way home without returning on their outgoing track. No doubt there is hereditary instinctive endowment, but there is teaching as well.

§ 15

The Story of the Otter

The Otter (*Lutra vulgaris*) is one of the most elusive of mammals, in great part nocturnal, shy of repeating itself or returning on its tracks, shifting in its hunting, and very thoroughly amphibious. It is much commoner in Britain than is generally supposed. Part of the secret of its survival we have already referred to—namely, the training which the mother gives to her offspring, but there is more. Thus it is always an advantage to have a catholic appetite, and while the otter depends mainly on fishes, it condescends to eat the mussels and limpets on the sea-shore and the frogs in the marsh; and, of course, it rises to wild duck and rabbit. Another feature of survival value is the otter's nomadism. In his fine study, *The Life Story of the Otter* (1915), Mr. Tregarthen calls it "the homeless hunter," "the Bedouin of the wild." It has been known to travel fifteen miles in a night, and not infrequently the holts where it lies up during the day are ten or twelve miles apart. It passes from tarn to stream, from river to shore; it swims out to an island in the sea; it explores the caves on the cliffs; it crosses the hills and hides in a cairn; it is always on the move—a gipsy among carnivores. In resourcefulness the otter is unsurpassed—lying hidden below the waterfall, wrenching a trap off under the roots of the alder-trees, diving at the flash of a gun, even hunting for pike beneath the ice of the lake. There are savage fights between two dog-otters who desire the same mate; the parents are often severely taxed to provide for the young; but the greater part of the otter's struggle for existence in Britain is in circumventing the difficulties of modern life.

The Common Hare

The Common Hare (*Lepus europæus*) might be called a gentle Ishmaelite. Everyone's hand is against it, but it is against no one unless it be greatly roused, for instance by a stoat approaching its leverets at play. Yet it extends all over Europe, except in Ireland, the north of Scandinavia, and the north of Russia. How does it survive? It seeks resting-places or "forms" from which it gets a good look-out over the surrounding country; it has long-sighted eyes, quicker ears, and keen smell; it utters a danger-call to its kin by grinding its teeth; its heart is such that it can put on full speed the moment flight is signalled; it rejoices in an uphill race; it criss-crosses its tracks so that even the astute fox is baffled; it disappears like an arrow when it is startled; and even when it is resting among the ferns and herbage, or on a ploughed field, it is almost invisible save as to its wide staring eyes. Much as it dislikes wetting its fur, which is slow to dry, it will swim across a broad river to baulk pursuit or to reach greatly appreciated dainties like musk and camomile. Epicure as it is, fond of tender corn and the sweet trefoil, of wild thyme and the

seashore pea, it has a long bill of fare, which always aids in survival, and it will pass from lichens on the rocks, which its cousin
the mountain-hare also eats, to the twigs of furze-bushes, and from
the leaves of dandelions to the fruits of the bramble.   Let us take
three more illustrations of the hare's astonishing fitness.   How
simple and yet effective is its habit of taking a great leap from
and into its "form" or nest, so that the scent track is interrupted.
In his fine study, *The Story of the Hare,* Mr. Tregarthen notes
that the doe leaves little scent when the young ones are helpless
in the nest, that is, about the month of April.   When a particular
nest is endangered, it may be by a hungry vixen, the doe hare will
transport its leverets to a safer place, carrying one at a time in her
mouth, at dead of night.   It is said that if the litter be over two—
cases of 4–6 are recorded—there may be a division into two nests!
Elusive is the word for a hare, but at the breeding season in
March the instinct of self-preservation wanes before sex-passion.
The bucks race about at a high speed in the open day and in the
open field, searching for the does and fighting with rivals.   They
box with their paws and kick with their hind-legs, and a common
trick is for one buck to jump over another, kicking back as he
does so.   The buck is a roving lover; he may consort with one doe
for a little while, but he soon seeks another.   The hare is a high-
strung creature, with quick-beating heart, rapid breathing, tremu-
lous ears, but it presents a brave front to persecution, now saving
itself by its alertness, and again by its capacity for lying low.   As
there is no burrow, it is not surprising to find that the leverets are
born furry and open-eyed, very different from the naked young
of their second cousins, the rabbits.

The Significance of Play

There are many playing mammals, and the work of Groos
in particular has shown that the play is of great importance in the
life of the creature.   Kittens chase a leaf whirled by the wind;
puppies indulge in a sort of sham hunt; young otters and stoats

are delightfully playful, and so are humble mammals like the water-shrews, which few people know much about. Lambs have many games, and goats have more; calves and foals have their races; leverets and squirrels their frolics. One may distinguish gambols, races, games like "tig," sham hunts, sham fights, and the endless game of "experimenting" in which monkeys are pre-eminent. Miss Romanes writes of her Capuchin Monkey: "He is very fond of upsetting things, but he always takes great care that they do not fall upon himself. Thus, he will pull a chair towards him till it is almost overbalanced; then he intently fixes his eyes on the top bar of the back; and, as he sees it coming over his way, darts from underneath, and watches the fall with great delight; and similarly, with heavier things. There is a washhand stand, for example, which he has upset several times, and always without hurting himself." This illustrates the game of experiment.

Similarly, Miss Frances Pitt records a game which two ravens in a yard used to play with a cat. One of the ravens, with a good deal of bluster, would make a frontal attack on the cat. This was met on the cat's part by the usual arching of the back and other expressions of contemptuous irritation. Meanwhile, however, the other raven approached quietly from behind and tweaked the cat's tail. Whereupon a rapid face-round, and the second phase of the game began, in which the ravens exchanged parts. There was no use in the performance; it was only a "ploy" in which the cat had its share.

What is the biological significance of the play of young mammals? It has been said that play is a good safety-valve for overflowing energy and exuberant spirits; it has been pointed out that motion is linked in a subtle way to emotion, and that pleased feelings naturally find expression in pleasant movements; it has been suggested that the playing period affords opportunity for trying new ways or exercising new gifts before the responsibilities of life become too stringent. There is good sense in each of these

Photo: W. S. Berridge.

EUROPEAN BADGER

The Badger or Brock (*Meles taxus*) is somewhat bear-like in its thick-set body, rounded back, short ears, depressed head, and flat-footed gait. It is over two feet in length, not counting the relatively short tail. It ranges through the northern parts of Europe and Asia, and still holds its own in some parts of Britain. Its survival may be traced in part to its vigour of constitution, to its burrowing and nocturnal habits, to its catholic appetite, and to its instruction of the young. The pair live together in the "earth," which is kept clean, although the repulsive smell does not suggest this. Badgers levy a slight tax on the eggs of game-birds, but they do very little harm, and it is a pity that they should be thoughtlessly eliminated.

Photo: Aberdeen University Museum.

THE HEDGEHOG

The Hedgehog (*Erinaceus europæus*) is an old-fashioned Insectivore, ranging from Ireland to the Ural Mountains. It survives in virtue of its nocturnal habits, its tough constitution, its armour of spines, its power of rolling itself up, and its capacity for hibernation. The spines are transformed stiffened hairs. The staple food consists of earthworms, slugs, and insects—both larval and adult; the back teeth bear mountain-top-like cusps well suited for dealing with this sort of diet. The pointed muzzle is adapted for probing into holes. The four to six young ones, born in a hedgerow nest or at the foot of a hollow tree, are at first very flat, with white soft spines pointing backwards, with a bluish skin, and without the power of rolling up.

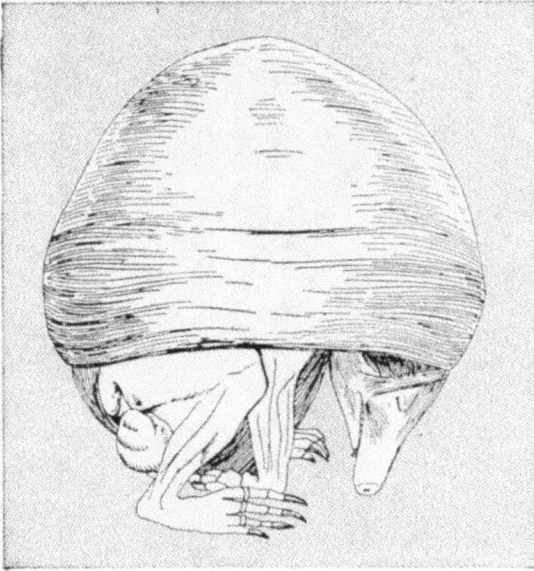

SKINNED HEDGEHOG

Showing the attitude of the animal when rolled up. Very noteworthy is the great dome of muscle which contracts the animal into a living ball. It will be noticed the muzzle is bent down very nearly to the toes, and that the fingers are touching the toes.

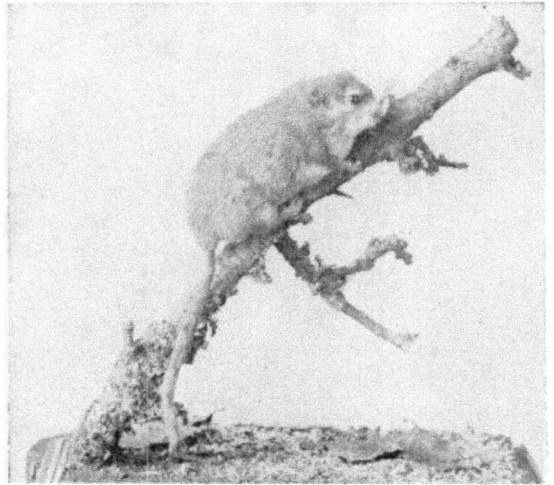

Photo: Aberdeen University Museum.

THE DORMOUSE

The Dormouse (*Muscardinus avellanarius*) is in some ways suggestive of a miniature squirrel. It climbs in the herbage and bushes; in the thicket in spring it makes a nest of grass and leaves. The tail is somewhat bushy; the eyes are relatively large. Head and body make up about 3 inches, the tail half-an-inch less; the colour is tawny brown above. The Dormouse frequents the central and southern districts of England. It accumulates fat in the summer, and is a light sleeper during the winter. If wakened too suddenly it is apt to die. If there be a second litter in autumn the young ones are said to die.

Photo: Royal Scottish Museum.

THE POLECAT AND ITS FAMILY

The Polecat (*Putorius putorius*) is also known as the fichet and the foumart (i.e. foul marten, because of its fœtid odour). It is much larger than the Stoat, with looser fur, darkish all over. It lives chiefly in wooded country, feeding on rabbits and birds, but is becoming very scarce in spite of its alertness and courage. A ferret is a domesticated form of the Polecat and is often an albino with no pigment in the hair or in the eyes (which look pink because of the blood shining through).

suggestions, but the most important idea is that the play period is the time for educating powers which are useful in after life. Play is the young form of work—a rehearsal without too great responsibilities, when mistakes can be made without too severe punishment. As Dr. Groos says, playing animals do not simply play because they are young; they continue young in order that they may play. In the course of ages playing instincts have been established in many mammals, and they make for success.

The Story of the Weasel

The Weasel (*Putorius vulgaris*) is one of the northern mammals common to Europe, Asia, and North America. It is a first cousin to the Stoat or Ermine, and an embodiment of virility. The spare sinuous body and the long neck suggest the snake, and the convergence simply means that the Weasel is adapted, like a snake, to making its way through narrow passages. The Weasel succeeds in virtue of a nimble brain, very keen senses, highly developed muscularity without any "spare flesh," and solicitous maternal care; but it would be unscientific to overlook its extraordinary courage. It will face up to a terrier, even to a man. It will leap up and catch a partridge already on the wing. "A pair will stand affectionately and nobly by each other in danger, and a weasel mother will defend her young to the last gasp." A weasel will explore a house and defy the house-cat; it will bluff a lot of roosting hens that could have pecked it to pieces; whining and daring, snarling and bristling, it will retrieve its young ones from under the feet of man.

§ 16

Social Mammals

Many mammals are gregarious and some go a step further and illustrate some measure of communal or corporate life. It is difficult to draw any hard-and-fast line. Gregariousness is illustrated by cattle, deer, wild horses, rabbits, kangaroos, and many more. The chief advantage is in the strength that numbers give

against an enemy. The members of the vegetarian herd trample
the carnivore to death. A small monkey attacked by an eagle has
no chance, but his cries bring a crowd of comrades to his aid, and
they may tear the bird of prey to pieces. Moreover, when there
is a herd, there is the possibility of having sentinels or outposts,
which warn the main body when danger draws near. The Rabbit
knocks loudly on the ground with its feet and the Marmot whistles
"danger." Whenever there is division of labour there is a sound-
ing of the social note. Thus when baboons are retreating the rear-
guard is formed by the old males, and Brehm tells the fine story
of the way in which they faced the dogs of his hunting party and
kept them at bay while the females retreated.

But one little monkey about half a year old had been left behind. It shrieked loudly as the dogs rushed towards it, but
succeeded in gaining the top of a rock before they had
arrived. Our dogs placed themselves cleverly, so as to cut
off its retreat, and we hoped that we might be able to catch it.
But that was not to be. Proudly and with dignity, without
hurrying in the least, or paying any heed to us, an old male
stepped down from the security of the rocks towards the
hard-pressed little one, walked towards the dogs without
betraying the slightest fear, held them in check with glances,
gestures, and quite intelligible sounds, slowly climbed the
rock, picked up the baby-monkey, and retreated with it, before we could reach the spot, and without the visibly discon-
certed dogs making the slightest attempt to prevent him.
While the patriarch of the troop performed this brave and
self-sacrificing deed, the other members, densely crowded on
the cliff, uttered sounds which I had never before heard from
baboons. Old and young, males and females, roared,
screeched, snarled, and bellowed all together, so that one
would have thought they were struggling with leopards or
other dangerous beasts. I learned later that this was the
baboons' battle-cry; it was intended obviously to intimidate
us and the dogs, possibly also to encourage the brave old
giant, who was running into such evident danger before
their eyes.

The Story of the Beaver

The Beaver is an aquatic mammal of a very different type, suited for rivers traversing wooded country. It is a thick-furred, plump creature, about 2½ feet long, with a flat, trowel-like scaly tail. It swims well with its webbed hind-feet and broad tail; it can remain about two minutes under water; it feeds mainly on bark. Its simplest home is a burrow with an entrance under water, but above the burrow there may be a surface pile of sticks, and from this rough-and-ready shelter there are gradations leading to a well-formed "beaver lodge" of sticks and grass, moss and mud. This includes a comfortable central chamber, with a "wood entrance" and a "beaver entrance." But the architecture varies with individuals and with the severity of the conditions of life. With more leisure, there is more art.

Beavers can cut down trees 10 inches in diameter; they use their chisel-edged incisor teeth, covered in front with orange-coloured enamel, to split off flakes of wood all round the base of the stem, but more towards the side nearer the water. The wind then brings the tree down, and the beaver's object is attained, namely, getting at the more palatable wood on the younger branches. These are cut into suitable lengths and stored in or near the lodge. The barked pieces may be added to the building. There is no doubt that beavers make dams of brushwood, stones, and mud, thereby securing a larger area for their wood-cutting and easier conditions of transport. It is likely enough that some of the dams were started naturally by floods which carried lodges and stores away and deposited them in shallow water; indeed, we can see the beginning of such a dam in many a river in wooded country. But the point is that the beavers strengthen, elaborate, and regulate what the river itself may have begun.

Even more remarkable is the digging of canals, by which the transport of the cut branches is made easier. They may be hundreds of feet long, and they are often about a yard broad and deep. They usually communicate between clumps of trees and the pond above the dam, but they may form a short cut between two loops of the river, or they may go right through an island. In the last case the work would not be justified until there was an open waterway from end to end. In some other cases a moist roadway between the pond and a pool in the wood might be gradually converted into a canal. Instances of "locks" have been recorded, but there is a tendency to forget that animals are more likely to take advantage of what exists or is hinted at in Nature than to discover new ideas or principles!

Beavers are notably gregarious, for there may be many lodges near a suitable wood. When there is overcrowding a migration occurs, the old houses being left to related new couples. Isolated males are often found, and some naturalists say that these have been expelled from the community for laziness or misbehaviour. There are no beavers left in Britain, but they flourish in Russia, in Siberia, and in Canada and other parts of North America. It is interesting to notice that in many places from which beavers have been gone for centuries, evidences of their work remain as "beaver-meadows" and the like.

Mutual Aid

Prince Kropotkin did a notable service in his book, *Mutual Aid, a Factor in Evolution* (1904), for he showed in a scholarly way the frequency of gregariousness, combination, co-operation, and sociality among animals. One answer-back that pays in the struggle for existence is to sharpen teeth and claws, i.e., to intensify competition; but another successful answer-back is to practise mutual aid. Even the individualistic carnivores may form packs as in the case of wolves and jackals; but there is more elaboration among the grazing herds. All kinds of beasts and birds of prey have proved powerless against the colonies of Russian sousliks. Combination gives strength to the sociable musk-rats of North America and to the prairie-dogs.

As far as the eye can embrace the prairie, it sees heaps of earth, and on each of them a prairie-dog stands, engaged in a lively conversation with its neighbours by means of short barkings. As soon as the approach of man is signalled, all plunge in a moment into their dwellings; all have disappeared as by enchantment. But if the danger is over, the little creatures soon reappear. Whole families come out of their galleries and indulge in play. The young ones scratch one another, they worry one another, and display their gracefulness while standing upright, and in the meantime the old ones keep watch. They go calling on one another, and the beaten footpaths which connect all their heaps testify to the frequency of their visits.

As Darwin said, "the individuals which took the greatest pleasure in society would best escape various dangers; while those that cared least for their comrades and lived solitary, would perish in great numbers." In short, the line of mutual aid is a trend of evolution, which has borne its finest fruits in mankind.

§ 17

Variety among Mammals

We see the March hares racing over the ploughed field, and the sloths creeping cautiously along the under side of the branches. The porpoises gambol in the sea, and the bats with erratic flight hawk insects in the air. The mole works its way for the most part underground, and the squirrel leaps adventurously from tree to tree. Whales are mammals of the open sea, and sometimes descend to great depths; monkeys are largely arboreal; antelopes are suited for the plains and the hippopotamus for the rivers. Wild cattle are gregarious, beavers are social, the sea-lion has his harem, the polar bear is solitary. We watch seals resting among the shore rocks, and bats hanging upside down from the rafters. In the winter the wolves join in packs, the stoat turns into the white ermine, the hedgehog sinks into hibernation. There are herbivores, insectivores, carnivores, specialists like the ant-eaters and the fish-eating seals, and others with a catholicity of appetite like badger and otter. A harvest-mouse only weighs about a halfpenny, an elephant's tusk may weigh 188 pounds. The Pigmy Shrew has a body under 2 inches in length, a whale may attain to 60 feet. A common shrew seems often to die in the year of its birth; an elephant may be more than a centenarian. But we need not go further; it is plain that there is extraordinary variety among mammals. This raises the question, what have they all in common?

General Characters of Mammals

Mammals are quadrupeds, except that the whales and sea-cows have lost all but vestiges of the hind-limbs, and perhaps another saving clause should be inserted for kangaroos, jerboas, and higher apes, which are more or less bipeds. In most mammals there is a distinct neck and a distinct tail, but the neck is practically obliterated in whales, and the tail is often much reduced (as in bear and rabbit) or practically absent (as in the higher apes).

Hairs are never entirely absent, for even in whales they are present in early stages of life and some, very richly innervated, often persist on the lips. The mammalian skin shows sweat-glands which get rid of surplus water and some waste-products, sebaceous glands which keep the fur sleek (absent in whales), and milk-glands which are normally functional in the females only.

In mammals only is there developed a midriff or diaphragm —a muscular sheet separating the chest cavity (containing heart and lungs) from the abdominal cavity (containing the stomach and other viscera). This midriff falls and rises in the breathing movements, and is of great importance in increasing and then decreasing the chest cavity, and thus helping the entrance and exit of air from the lungs.

Mammals have many skeletal peculiarities which separate them off from all other backboned animals. The vertebræ (back-boned-bodies) and the long bones have terminal caps which ossify

Photo: Riley Fortune, Harrogate.

THE OTTER (*Lutra vulgaris*)

This member of the Bear tribe of Carnivores is about 2 feet long, with 16 inches more to the strong tail—which helps in swimming. The fur is thick and soft, deep brown above. The claws are of use in burrowing, but the hands and feet are likewise webbed for swimming. In its present geographical distribution the Common Otter extends from Ireland to India. It holds its own in virtue of strength, strong claws and teeth, keen senses, alert wits, roving habits, versatility of diet and hunting-grounds, resourcefulness when hard pressed, ability to lie for a long time hidden under a bank, and careful education of the young.

Photo: From Royal Scottish Museum.

VIXEN AND HER PLAYFUL CUBS

The Fox (*Canis vulpes*) is the only wild member of the Dog tribe of Carnivores now left in Britain. There is considerable variation in size and colour in different parts of the country. Foxes associate in pairs, and the four to seven cubs remain for a considerable time under the care and tuition of the vixen. The young ones are very playful and enterprising. Foxes make "earths" or burrows in hills and woodsides, and most of the day is spent in hiding. They come out at dusk and hunt for small mammals and birds, and a variety of creatures of lower degree—down to shellfish on the shore.

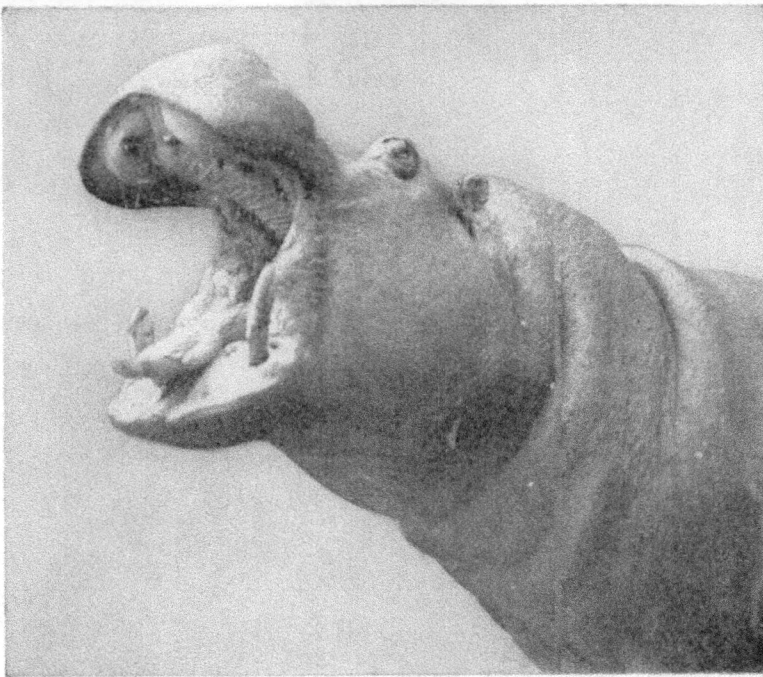

HIPPOPOTAMUS

The Common Hippopotamus (*H. amphibius*) of Africa is one of the modern giants (4 tons in weight, 14 feet long), but there is a dwarf species in Liberia. Their nearest relatives are the Pigs. The huge creature can swim with efficiency, and occasionally "puts out to sea"; it can also walk along the bed of the river, remaining immersed for ten minutes. It is vegetarian. The body is almost hairless. The nostrils are situated high up, as is suitable in an aquatic creature. There is a strange bloody sweat. The hippopotamus is the Behemoth of the Book of Job—"the chief of the ways of God."

HARVEST-MOUSE ON THISTLE

The Harvest-Mouse (*Mus minutus*) is, next to the Pigmy Shrew, the smallest of British mammals, weighing only about one-fifth of an ounce. It can run up a stem of wheat, and in its descent it uses its tail in a monkeyish fashion. A nest of coarse grass, with a side entrance, is built between three or four stalks of corn, and there five to nine young ones are born—blind and helpless, but developing rapidly.

apart from the main part of the bone; the surfaces of the vertebræ are usually flat or gently rounded; with four exceptions there are seven neck vertebræ—whether it be in the long straight neck of the giraffe or the compressed inconspicuous neck of the whale; the lower jaw is one bone on each side and works on a bone of the skull called the squamosal; the skull moves by two knobs (or condyles) on the first vertebræ, whereas birds and reptiles have only one condyle; the drum of the ear is connected with the internal organ of hearing by a beautiful chain of three small bones—the hammer, the anvil, and the stirrup—by which the vibrations are conveyed inwards; there is a complete bony palate separating the mouth-cavity from the nasal passage above; almost without exception there are two sets of teeth in sockets; except in the oviparous mammals the bone of the shoulder-girdle called the coracoid, which is very strongly developed in flying birds and in reptiles, is represented merely by a small process of the shoulder-blade or scapula.

The cerebral hemispheres of the fore-brain are much more developed than in other vertebrates, and their surface is very generally covered with convolutions (see figure facing p. 157, Vol. I). The heart is four-chambered; the temperature of the blood remains in most cases practically constant; the red blood-cells are circular discs (except in camels, where they are elliptical in outline as in other Vertebrates), and the nucleus of the mammalian red blood corpuscle disappears as the corpuscle develops; the lungs lie freely in the chest cavity (they are fixed in birds), and inspiration is the active process (the opposite in birds); the vocal cords are at the top of the windpipe (at the foot in birds); the egg-cells are very small except in the egg-laying forms, and, with the same exception, the young are born viviparously, i.e., as living well-formed young ones, which are for a while nourished on milk. This enumeration of salient characters is indispensable if we are to understand how this class of Mammals stands apart from the other great classes of backboned animals, namely, Birds, Reptiles, Amphibians, and Fishes. Aristotle knew that a whale is not a fish; unless we understand the general features of mammals we will not appreciate his insight.

The dominantly successful orders of present-day mammals are: (1) the Carnivores (cats, dogs, bears, seals, etc.); (2) the hoofed Ungulates (horses, tapirs, rhinoceros, cattle, pigs, hippopotamus, camels, perhaps also including the elephants); and (3) the monkeys and apes or Primates. These represent three great lines of evolution. On the Carnivore line the premium is on teeth and claws, quick senses and alert movements. On the Ungulate line the premium is on swiftness, on the power of covering long distances in search of herbage, and on such weapons as horns and hoofs, rendered more effective still when their possessors are gregarious. On the Primate line the premium is on the power of climbing, the emancipated hand, and the restless brain. Below the level of true Primates are the Lemurs, or half-monkeys— ghost-like nocturnal creatures, mostly confined to the forests of Africa and Madagascar.

Not very far off the Carnivore line of evolution, but much more primitive, is that of the Insectivores (e.g. moles, hedgehogs, and shrews); and the Bats with their power of flight must be regarded as the specialised descendants of arboreal Insectivora.

Balancing the Insectivores there is the somewhat humble order of Rodents, on a quite different evolutionary tack, the rats and mice, squirrels and porcupines, rabbits and hares. The toothed Whales and baleen Whales are mammals that have taken secondarily to marine life, and are as specialised for swimming and diving as bats are for flight. And besides these well-known orders, there are the sea-cows or Sirenians (including nowadays two genera only, the dugong and the manatee), the old-fashioned Edentates (sloths and armadillos, ant-eaters and pangolins), perhaps to some extent survivors of the archaic mammals. But more primitive in their affinities than all these are the Marsupials (mostly confined to Australia); and lowest of all are the egg-laying Monotremes, also Australasian.

BIBLIOGRAPHY

BEDDARD, *Mammals,* vol. x. of *Cambridge Natural History* (1902).
FLOWER AND LYDEKKER, *Mammals, Living and Extinct* (1891).
HUTCHINSON, *Extinct Monsters* (1893).
INGERSOLL, *The Life of Mammals.*
JOHNSTON, *British Mammals* (1903).
LANKESTER, *Extinct Animals* (1909).
LYDEKKER, *British Mammals* (1896).
NELSON, *Wild Animals of North America.*

XIV
NATURAL HISTORY

NATURAL HISTORY

III. THE INSECT WORLD

Insects almost Ubiquitous

THE immense and varied group of Insects constitutes by far the largest class in the Animal Kingdom; it numbers as many as 200,000 named species, the majority of which are predominantly active types. Such a wealth of forms—the species in a single family of insects may outnumber the stars one can count on a clear night—shows that, as a class, Insects are extraordinarily successful. Many reasons are given for this dominance, all pointing to the striking fact that insects, by means of manifold adaptations, are able to fill many niches, and so attain a wide distribution. Few haunts are destitute of insect life. Butterflies and mosquitoes are known to penetrate to extreme Arctic regions; a small kind of butterfly is found in Ecuador at an elevation of 16,500 feet; insects inhabit desert tracts far out of reach of water; and limestone caverns have their cave-dwellers, often pale and blind unless their descent to this unusual haunt has been comparatively recent. Many forms live in fresh water; even hot-springs have their insects, and some beetles, for instance, are found on the tidal zone of the sea-shore. The actual sea seems very unsuitable for insect life, and yet there is a family of Skimmers (Halobatidæ) which run about on the surface of the open ocean, and even dive when it is stormy.

The Success of Insects

Insects are typically winged creatures, and their power of flight extends their range, giving the opportunity to colonise new areas and to migrate to fresh localities in times of stress. Their bodies are extremely well adapted from the mechanical point of view; their sense-organs are highly developed—sensitive feelers, compound eyes, and so on—and their mouth-parts are remarkably adapted to suit different modes of feeding. Probably much of their success in the struggle for existence is due to the adaptations of their circulatory and respiratory systems, which enable the nutrition of the organs of the body to go on with great rapidity. The tissues are continually bathed in nutritive fluid, while every part of the body is kept aerated by the extensive system of air-tubes. These facts account for the abundant energy and consequent activity which is so characteristic of the class. It may be doubted if the insect's blood ever becomes impure. Another factor tending towards success is the change of habit due to the change of form which occurs during the course of many life-histories. This implies changes in diet, and therefore a lessening of the drain on any particular foodstuff. In other ways, also, the changes of form and habit may lead to survival in the struggle for life, for there is frequently a tiding over of difficult times; for instance, quiescence during periods when conditions of temperature and food are unfavourable. Many insects pass the winter in a lethargic state inside well-protected cocoons.

Protective Adaptations

Another factor which helps to give success to insects in maintaining their hold in various habitats is the way in which general form and colour are adapted to the environment. Protective colouring in animals has formed the subject of a special article, but it may be noted that there are no clearer instances of *protective resemblance* than among insects. Not only do they very often closely resemble the general colour of their natural surroundings, but form, as well as colour, may add still more to this similarity, which gives security to the insect by concealing it effectively from its enemies. We can thoroughly understand the wonder of this protective resemblance only when we study it under natural conditions; many very gaudy butterflies can hardly be distinguished from flowers when they alight on plants. Many moths in their resting position hide the bright colours of the hind pair of wings with the duller fore-wings, which may nearly resemble lichen or the bark of trees.

The coloration may afford an effective protection in other ways, by Warning and by Mimicry. Some insects, such as the Wasp or the Lady-Bird beetle, positively court attention with their vivid colouring and markings; they are coloured, not to be hidden, but to be seen. Such insects always have some other form of protection, a sting or an unpleasant taste, which their enemies come to associate with their striking hues and therefore avoid. No doubt conspicuous individuals will be snapped at and killed while birds and other enemies are experimenting, but the enemies learn by experience, and the species with the warning colours gradually attain a position of security.

§ 1

Pedigree

The pedigree of Insects is obscure. They belong to the large group of the joint-legged Arthropods, which shows numerous affinities with the ringed worms or Annelids, but also many advances such as the greater development of appendages. In Peripatus and its allies, which are widely distributed over the world, worm-like, velvet-skinned little creatures, shy and nocturnal in habit, we find living links between Annelids and Insects. In their excretory tubes, muscular arrangement, and hollow appendages they strongly suggest the ringed-worm type, but they combine with these and other Annelid features distinct indications of Arthropod characters, such as the system of breathing tubes and the appendages in the service of the mouth, which reach fuller development in the class of Insects.

General Characters of Insects

Insects, Peripatus, Centipedes, and Millipedes have in common a respiratory system consisting of tubular tracheæ, which marks them off from the gill-breathing Arthropods (Crustaceans), and sensitive feelers, which distinguish them from the Spider and Scorpion group (Arachnids). In the class of Insects the body in the adult state is divided into three main regions: (1) the head; (2) the thorax or fore-body; (3) the abdomen or hind-body.

The outer covering of most insects is hard and firm, composed of a non-living cuticle made of chitin, a somewhat horn-like substance secreted by the underlying living skin. The chitinous plates, which make a protective armour, are firmly fused in the head region, but in the thorax and in the abdominal part the different rings are joined by flexible areas, permitting more freedom of movement. Thus the segmented architecture of the body is more clearly seen in the thorax and abdomen than in the head region, where fusion has obliterated the boundaries of the successive segments of the body. In rapidly flying insects there is often a fusion of thorax rings to form a firm basis for the action of the wings.

It must be clearly understood that in the insect's body the muscles are *inside the skeleton,* whereas in ourselves the skeleton is covered by the muscles. The two plans of architecture are utterly different.

The Insect's Head

The insect's head, which bears one pair of feelers or antennæ and usually three pairs of jaws, is relatively small, firm, and compact, separated from the thorax by a narrow membranous neck allowing freedom of movement. One sees this very well on a common house-fly. All adult insects (except some primitive and some degenerate species) have a pair of compound eyes, though simple eyes may be present also. The compound

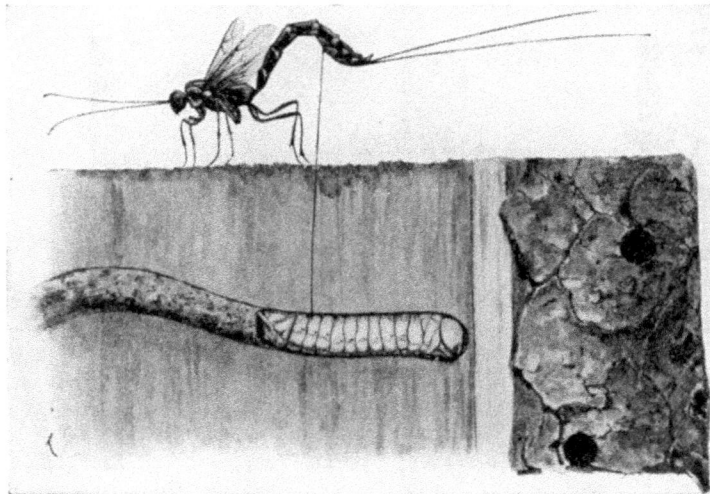

THE FEMALE OF AN ICHNEUMON-FLY (*Rhyssa persuasoria*) BORING WITH ITS OVIPOSITOR IN A FELLED TREE AND LAYING ITS EGGS IN THE GRUB OF THE GIANT SAW-FLY (*Sirex gigas*), WHICH IS DEEPLY EMBEDDED IN THE WOOD

The eggs of the Ichneumon-fly hatch inside the grub of the Saw-fly, which is then devoured by the Ichneumon grubs.

Photos: J. J. Ward.

FIG. 1.—Dragon-fly (*Vibellula depressa*) emerging from its nymph skin. "Nymph" is the name given to the immature stage of insects whose life-history shows *incomplete* metamorphosis. There is no pupa. The adult stage is reached at the final moulting of the nymph skin. (See illustrations Nos. 2 to 6.)

FIG. 2.—At the end of three minutes the Dragon-fly is free, all but the tip of its abdomen.

FIG. 3.—A moment later it has extricated its tail end and is free. Its wings then commence to extend from their folds. (See Fig. 4.)

FIG. 4.—Five minutes after its emergence its wings are fully extended.

FIG. 5.—After the wings have dried and come under muscular control, the insect raises itself and brings them into the natural resting attitude, as shown in Fig. 6.

FIG. 6.—Short-bodied Dragon-fly just expanding its wings after emergence from the nymph skin which is seen near its head.

eyes project on each side of the head as convex, immovable structures. There is only one pair, though each eye may be partially divided, as in some of the aquatic Whirligig Beetles in which half of the eye is directed up to keep a look-out for danger from above, while the other half is scanning the water below in search of prey. The compound eye consists of a great many similar parts—each a complete organ of vision but requiring the surrounding elements to form the whole image. Each of the many elements of the eye makes a small image, so that the whole image is a mosaic of separate contributions, which combine in a unified visual impression conveyed to the brain. For the amorous insect does not see 1,000 desired mates,—one through each of its eye-elements. The question is not an easy one, but it should be noticed that, in some cases, e.g., fireflies, the eye-elements no longer act separately, but a single combined image is thrown on the back of the eye. (See figures facing pages 307-308.)

The antennæ are appendages set in sockets on the crown of the head, and consist of a series of joints, varying from one or two to a large number, and of many different shapes. They are of the greatest importance to the insect as organs of touch, by means of sensory bristles connected with underlying nerve-fibres, and also in connection with the sense of smell. Of hearing, in insects, very little is securely known. Further, the head carries three pairs of mouth-appendages (homologous with legs), which are variously transformed for different modes of feeding, biting, or sucking. It is very interesting to find that the same three parts are changed in scores of different ways.

Insects' Legs

The legs, which are borne on the three rings or segments of the thorax region, show many different peculiarities to suit different habits. The front pair is considerably lengthened in certain beetles that climb about the bark of trees; in the Mole Cricket they are converted into burrowing implements, the terminal joints being arranged as shears for cutting through plant roots (see figure facing page 312). The "Praying Mantis" and the Water-Scorpion both show the fore-legs modified into pincer-like traps for seizing prey. Usually the middle pair is not greatly modified, but in some water-bugs, like the Water-boatman, the middle legs are the longest and have become effective oars for rowing on the surface of the water. The hind pair of legs of many insects is elongated for jumping, as in grasshoppers and locusts and some beetles. Certain beetles and bees and wasps have a "comb" or bristle-lined cavity on the leg by means of which they clean their feelers, while some butterflies use their feeble front legs to brush off dust from their heads. Ants are particular about their toilet. In the course of the day's work an ant's antennæ may become soiled. On its first pair of legs it is provided with what we may call brushes and combs, as we have described, and the ant may be seen to draw its besmeared antennæ through this brush-and-comb arrangement on the fore-legs. One of the legs will be passed over its head and body, its other legs sweeping off every particle of "dirt." No cat is more fastidious over its toilet. Ants will even wash and brush each other, just as they will exchange greetings, as they meet, by movements of their antennæ. The hind-legs of bees show a modification for pollen-gathering, a broadening of the "shin" to make a "basket," into which the pollen is swept by special bristles.

Insect's Breathing

Breathing takes place by means of a system of air-tubes or tracheæ which penetrate to every hole and corner of the body. Tracheæ arise as inpushings of the skin, and the layer of chitin which lines them is continuous with that which covers the whole body. In the larger air-tubes this chitin is thickened spirally in threads, and this keeps the tubes from collapsing. Air enters the body by openings (spiracles or stigmata) occurring on most of the body-rings.

Through these spiracles the air is driven out by movements of contraction; fresh air passes in passively as the body expands. As in birds, so in insects, expiration is the active part of the breathing process. The air-tubes fork and re-fork, sending side-branches to every corner of the body, even to the tips of the feelers, so that the whole body is thoroughly aerated. The extensiveness of the air-tube system compensates for the relatively poor blood-system. In aquatic forms various devices are adopted to secure a supply of oxygen. Some water-insects come to the surface to breathe, others, like young may-flies, have special structures—tracheal gills—of different types. The Water-Beetle (Dytiscus) has its spiracles on its back, and when it dives under water it carries with it, in an air-tight compartment between its back and its hard wing-covers, enough air to last for several minutes. The bubble-of-air method is another plan, adopted by Whirligig Beetles and some water-bugs, whose covering of fine hairs entraps bubbles of air, ensuring a sufficient supply of air about the body for a short time under water.

In addition to the respiratory system there are inside the body of the insect all the usual organs—food-canal and associated parts, a heart, excretory organs, reproductive organs, and so on. Some insects are so small that they can creep through the eye of a needle, and it is difficult to believe that in such minute dimensions all the ordinary organs are packed away.

Locomotion

Insects are essentially active, and they exhibit various kinds of locomotion. Many grubs and maggots are quite passive, but even limbless larvæ, though naturally not so active as the legged types, have their ways of getting about. They may jerk themselves along with the aid of bristles or jaws, they may make relatively enormous leaps into the air by taking their tails in their mouths and suddenly letting go, or they may swing themselves from place to place by paying out silken lines from their mouths. Young dragonflies propel themselves through the water by means of the forcible expulsion of water from the end of the food-canal. Insects walk, run, and jump with the quadrupeds, fly with the birds, glide with the serpents, and swim with the fishes. It is often asked how a fly contrives to walk up smooth, perpendicular surfaces, and one answer is that a vacuum is made below a little soft pad which is present on the foot. Another explanation is, that there seems to be a slight exudation of adhesive moisture from the foot. Beetles, which have relatively strong legs, very different from the weak legs of a butterfly, can run with considerable speed, while many insects—one has only to think of a flea or a grasshopper—are pre-eminently leapers. The most primitive insects, the spring-tails and bristle-tails, are entirely wingless, but a spring-tail is an expert jumper. It has at the end of the body an effective leaping apparatus consisting of two elongated prongs, which are bent under the abdomen and pressed down, affording such a leverage when the retaining "catch" is released that the insect springs forward a relatively long distance compared with the size of its body.

From great leaps to the beginnings of flight is an understandable step in progress, and most insects are fliers. There are many patterns of wings, but essentially they are lightly built, mere flattened sacs of skin, often transparent and fragile, but beating the air with an extraordinarily rapid motion. It has been calculated that a fly makes 330 wing-strokes in a second, a humble-bee 240, a wasp 110, a dragon-fly 28, and a butterfly 9. The rapidity of the movement produces a hum or buzz. Bees and wasps have two pairs of membranous wings, but the fore-wing and the hind-wing on each side act as a single organ, for the hind-wing has a row of minute hooklets which fit into the curled-over posterior edge of the fore-wing and lock the two wings together. In dragon-flies the two wings are not attached, but the two pairs are co-ordinated by the action of very strong muscles, and the larger dragon-flies are excellent fliers. They are prob-

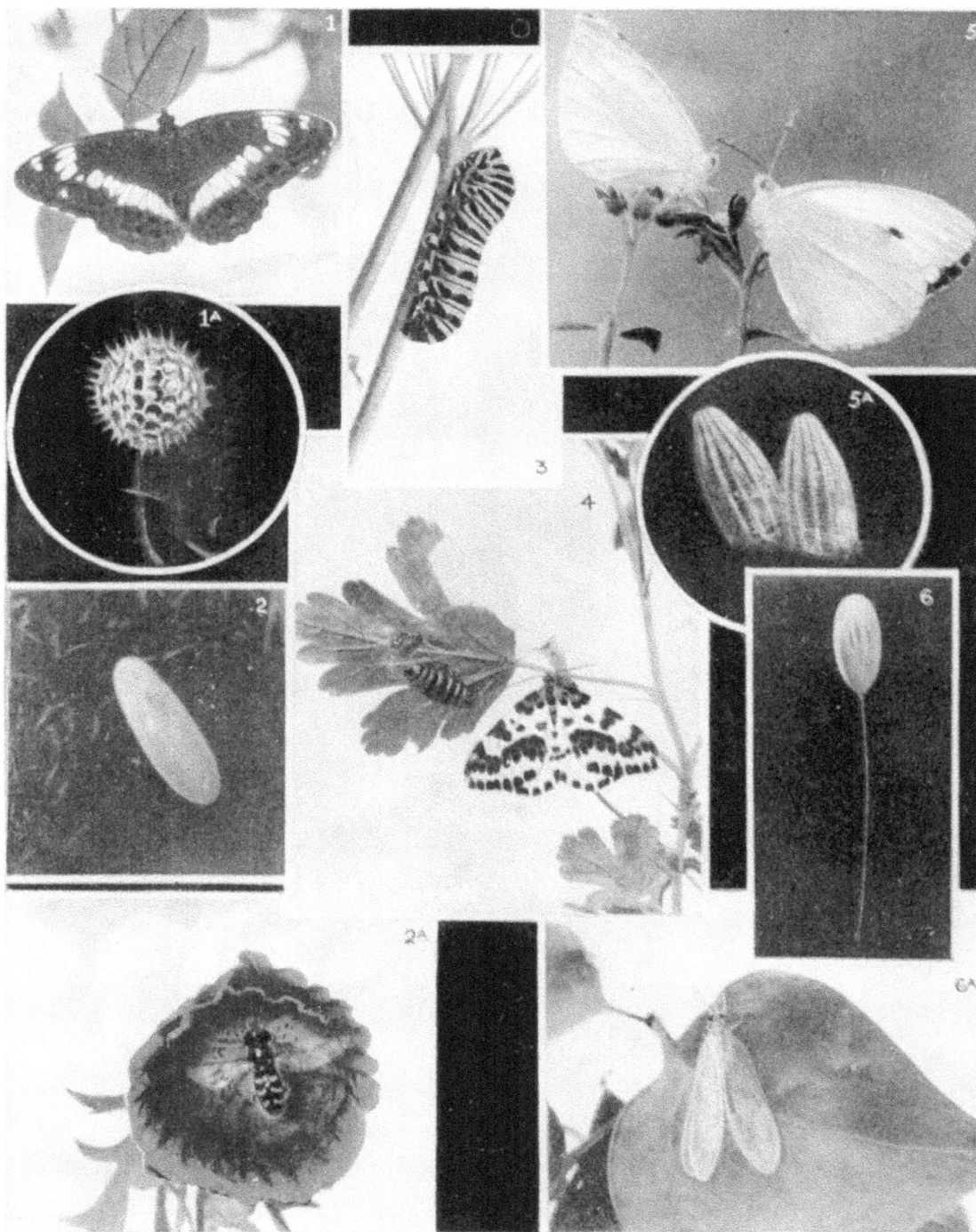

Chromo-photos: J. J. Ward, F.E.S.

INSECT LIFE

1, White Admiral Butterfly (*Limenitis sibylla*), and 1A, its egg (magnified 25 diameters); 2A, Hover Fly (*Catabombap yrastri*) on Poppy flower and, 2, its egg (magnified 30 diameters); 3, Full-fed larva of the Swallow-tail Butterfly (*Papilio machaon*); 4, the Magpie or Currant Moth (*Abraxas grossulariata*) just emerged from its pupa skin—shown on the leaf; 5, Large White Butterflies (*Pieris brassicæ*) love-making, and 5A, eggs of the Large White Butterfly (magnified 25 diameters), which are deposited on plants of the cabbage family; 6A the Lacewing Fly (*Chrysopa vulgaris*) and 6, one of its stalked eggs (magnified 3 diameters).

ably helped in steering by the weight of their bodies, the lightness of most insects being against good steering as they are liable to be blown about by the wind.

Whatever the pattern of wing or the speed of the wingbeats, the total distance insects can fly is not great; they seldom wander far afield. Some insects literally fly but once. A mayfly may rise at noon from the water that cradled it, and by sundown its aerial dance of love may be over and its lifeless body be floating on the surface of the pool.

§ 2

Instincts and Intelligence

Insects are largely creatures of instinct, with inborn capacities for doing apparently clever things, but yet with some degree of intelligence. In an animal's behaviour there is often, no doubt, a mingling of different kinds of activities unified in a way that baffles analysis. In many cases their behaviour under new conditions, their powers of effectively meeting new ends, go beyond mere instinct. What are we to say of the following?

The tailor-ants, common in warm countries, make a shelter by drawing leaves together, and their co-operative hauling is admirable; their mandibles are their needles, if you like, but they have nothing to fix the leaves with; what does each do but take a larva in its mouth so that the silk secreted from the offspring serves as adhesive gum.

The tailor-ants nest in trees, and they sometimes find it difficult to bring two rather distant leaves close enough together to be sewn. Then, as Bugnion relates, they have recourse to a perfectly extraordinary co-operation. Five or six will form a living chain to bridge the gap. The waist of A is gripped in the mandibles of B, who is in turn gripped by C, and so on—a notable gymnastic feat. Time does not appear to be of much account, but they work definitely towards a result, and many chains may work together for hours on end trying to draw two leaves close to one another.   We could not have a better instance of social co-operation.

An eye-witness, Mr. L. G. Gilpin-Brown, writes from Ceylon:

> Sometimes one will see an ant with a larva on its mandibles stalking aimlessly about on the outside of the nest. It stumbles on a small hole. It proceeds to study that hole, walks all round it, walks over it, and eventually decides that it really *is* a hole, whereupon it proceeds to business. Feeling around the edge with its antennæ it dumps the head of the larva on one side so as to fasten the thread of silk there, moves over and fastens it down on the other side, comes back again, and so on; each trip leaving a thread of silk behind until the hole is completely sealed up.

A common harvesting ant of South Europe collects seeds of clover-like plants, lets them begin to sprout so that the tough envelopes are burst, exposes them in the sun so that the germination does not go too far, takes them back underground and chews them into dough, and finally makes this into little biscuits which are dried in the sun and stored for winter use. What a brilliant idea—and yet it cannot be that!—is suggested by the semi-domestication of green flies by a certain species of ants! and what shall we say of the slaves which others bluff into service? Many white ants or termites grow highly nutritious moulds in extensive, specially constructed beds of chewed wood, and some of the true ants show a similar habit.

On wayside plants in early summer we see everywhere the frothy masses called cuckoo-spit, each made by a larval frog-hopper which whips a little sugary sap, a little ferment, and a little wax into a strange persistent foam, protective against enemies and against the heat of the sun, the creature literally saving its life by blowing soap-bubbles. Not far off, on a bare sandy patch, are the deep shafts sunk by the grubs of the beautiful green Tiger Beetle. The grub, with quaint somersault movements inside the shaft,

THE CATERPILLAR LARVA (SECOND STAGE OF LIFE-
HISTORY) OF THE DEATH'S-HEAD HAWK-
MOTH. THE CATERPILLAR IS FEEDING ON
POTATO VINES

PUPA OF DEATH'S-HEAD HAWK-MOTH (MALE)

The larva (a caterpillar in this case) ceases to be active and
passes into a state of quiescence, the pupal stage, during which
the body is undone and rebuilt. Out of the pupal case the
fully-formed insect emerges.

THE DEATH'S-HEAD HAWK-MOTH

The black spotted yellow patch upon the thorax gives the impression of a human skull.

DRIVER-ANTS ATTACKING A SNAKE

The Horned Viper, shown in the illustration, was attacked whilst casting its skin. The ants covered every portion of its body, hanging on by their pincer-like jaws. The snake writhed and struggled for a quarter of an hour, but in the end was killed and eaten by the ants.

thrusts the loose earth with great force into the walls, and beats them smooth. Eventually it fixes itself near the top of the shaft so that the roof of its head forms a trap-door. When an ant or some other small insect settles down on this living lid, the grub suddenly explodes like a jack-in-the-box, hurling its victim violently against the hard upper edge of the shaft-wall. The sucked body is afterwards jerked out. The world is full of these inventions.

How are we to understand the behaviour of one of the Digger Wasps which lays its eggs in a sunk shaft, and provisions this with paralysed caterpillars? While the hunting and storing are in progress, the wasp shuts the mouth of the shaft after each visit, but does so in a rough-and-ready fashion. When the larder is full, however, it seals the entrance with earth and makes a neat job of it; nay, it takes a minute pebble in its jaws and beats the earth smooth. Who said animals could not use tools? It seems that using the pebble is not part of the instinctive routine, but is an individual touch, probably with more vivid awareness than is associated with the rest of the agency. But the difficulty is to think of the origin of either the routine or the finishing touch without postulating intelligence, or, at least, some appreciation of significance.[1]

Homing

It is well known that ants and bees can find their way home from a distance. Ants evidently take impressions, by touch, sight, or sense of smell, of certain signposts. There may even be a "muscular memory" of the movements effected and of the amount of work done. Probably ants improve gradually in their way-finding as they learn to make use of a combination of the various hints. An interesting experiment suggested that bees build up a knowledge of the country round about the hive. Professor Yung of Geneva took twenty bees from a hive near the lake and liberated them at a distance of six kilometres in the country. Seventeen returned to the hive, some within an hour.

[1] J. Arthur Thomson, *Secrets of Animal Life*

Next day the successful seventeen were taken on a boat to a distance of three kilometres on the lake. When liberated they flew off in all directions, but apparently they missed the necessary signposts, for none of them found their way home. On the other hand, experiments have given results that indicate that bees have a "sense of direction," comparable to that of carrier-pigeons. Even bees with their eyes obscured have been known to make a "bee-line" for the hive from considerable distances. But there is no doubt that bees make cautious and systematical trial "flights of orientation" when a hive is placed in a new position.

Intelligent Behaviour

An outstanding feature of Ants is that of instinctive socialisation. They do not live unto themselves, but for the general good of the community. They are indefatigable, but whether they toil consciously for the sake of anything, or what we are to read into their capacity for unified action, who shall say?

> It is difficult to accept the opinion of some naturalists that instinctive behaviour is unaccompanied by any awareness of meaning or feeling of the end. Whenever this difficulty is obvious, it is customary to say that intelligence has for the time being taken the reins. In any case, the *facts* are wonderful enough.

It is among the Social Insects that the most pronounced evidences of intelligence are found.

> Intelligence is an eminently social faculty [as Kropotkin says]. Language, imitation, and accumulated experience are so many elements of growing intelligence of which the unsociable animal is deprived. Therefore we find, at the top of each class of animals, the ants, the parrots, and the monkeys, all combining the greatest sociability with the highest development of intelligence. The fittest are thus the most sociable animals, and sociability appears as the

chief factor of evolution, both directly, by securing the well-being of the species while diminishing the waste of energy, and indirectly, by favouring the growth of intelligence.

Mutual help is practised extensively among insects of various kinds. The Burying Beetles, which usually lead a solitary life, call to their aid a number of their fellows when there is a corpse to be buried. Many caterpillars weave a silken web to make a shelter for a whole brood, while the full-grown Procession Caterpillars march together from their feeding-ground on the trees to a soft place on the ground where they can bury themselves and become moths. Locusts display gregarious habits also which are of mutual advantage; for instance, it is a common practice for the wingless young to make a living bridge over a moderately broad stream, plunging into the water and grappling for sticks and straws, and scrambling for a breathing space on their comrades' bodies, till the whole swarm passes across the stream. Comparatively few are drowned, as the same individuals are seldom in the water the whole time. Such associations for mutual aid suggest the beginnings of societies, but they are not nearly so highly evolved as those seen among the termites, ants, bees, and wasps, where the social habits extend to the welfare of the young, and co-operation reaches a high level. Kropotkin says, "If we knew no other facts from animal life than what we know about the ants and the termites, we already might safely conclude that mutual aid (which leads to mutual confidence, the first condition of courage) and individual initiative (the first condition for intellectual progress) are two factors infinitely more important than mutual struggle in the evolution of the animal kingdom." The fact is that in the struggle for existence, which includes all the answers-back that living creatures make to environing difficulties and limitations, sociality pays just as well as intensified competition, or, it may be, *pays better*.

§ 3

THE STORY OF ANTS

The Marvels of the Ant-hill

Of all insects, Ants must be placed on the highest level, for none have better mastered the art of living together in a mutually beneficial manner, and many ant communities show considerable elaboration. Let us, then, take the case of ants as a particular illustration of the distinctive features of insect societies. Here we have "a community of separate individuals with more or less of a corporate life, and with the power of acting as a unity." Many Ants live for a number of years, so that one generation may teach another the profitable arts which lead to the success of the community. The welfare of the species is the important matter, and the individual is often sacrificed, as well as specialised, for the common good. There are three types of individuals—winged males, winged females, and wingless "workers" or undeveloped females; and the workers may be of different kinds, large and small—or with huge mandibles in the "soldier" type. We see a division of labour. The busy workers tread the neighbourhood of the nest into a pattern of "ant roads" by which they come and go on their foraging expeditions. Smell counts for much in way-finding. Within the nest, the workers have their home duties, they look after the young, feeding them and carrying them from room to room to secure a suitable temperature, and they bite open the cocoons when the perfect insects are ready to emerge.

Mutual aid and harmony seem to reign within the community, but there are terrible wars with other species, which are carried out in a well-organised fashion. Ants have the instinct of acting together and seldom make individual attacks, but they never seem to hesitate to sacrifice themselves for the protection of the community. Sometimes these warlike expeditions are initiated with a definite end in view, that of capturing slaves.

For instance, the Amazon Ants, which have jaws well suited for warfare but inconvenient for the peaceful occupations of life, habitually keep slaves to wait upon them. Professor Wheeler thus describes them: "While in the home nest they sit about in stolid idleness or pass the long hours begging the slaves for food, or cleaning themselves and burnishing their ruddy armour, but when outside the nest they display a dazzling courage and capacity for concerted action." Scouts report their discovery of a Brown Ant colony, and a raid promptly follows, the Amazons returning victorious with a large number of prisoners, which become faithful slaves. Darwin's suggestion "of the origin of slave-making was that many ants capture the pupæ of other ants for food, that some of the stored pupæ might be unintentionally reared, that if their presence in the community was not resented but proved useful, the slave-making habit might make ground."

Like the Termites, the true Ants frequently have guests within their homes. Certain little crickets find shelter and abundant food in this hospitable haunt. They beg food from the ants, and usually they shamelessly steal from the newly-fed young ants. Beetles, too, with a peculiar fragrance that makes them welcome guests, persuade the ants to share the sweet substances they carry in their crops, by stroking them till they deliver up the coveted dainty. One species of ant carries mites about on the body, feeding them and caring for them, but apparently deriving no benefit from them. Evidently ants are fond of keeping pets!

One of the peaceful occupations ants pursue is keeping "cattle." Their "cows" are little aphides or green-flies, which they cherish for the sake of the sweet "honey-dew" that exudes from their bodies. Possibly at first it was simply a matter of feeding at the same table, when the ants would discover the sugary fluid and get into the way of licking the green-flies. The eggs of a certain aphis, which are of no direct use to the ants, are brought into the nests and protected carefully from the severities of winter until the warm weather comes, when the young aphides are brought out and put on their food-plant, walled in by little "cattle-pens" of earth. By keeping these eggs safe for six months the ants ensure a supply of the food delicacy during the following summer—a truly remarkable case of prudence!

The Wonderful Leaf-cutting Ant

In North America there are "agricultural ants" which weed a space near the nest and only allow plants with edible seeds to grow there. These seeds they gather in due season, and store in the form of little biscuits which are made from a chewed-seed dough dried in the sun. Another industry is the cultivation of fungi for food—another point in which they agree with the Termites—and this habit is seen among the Leaf-cutting Ants. The fungus is grown in the underground nest on a spongy framework of chewed leaves, and the ants not only keep undesirable fungi from growing amongst their peculiar delicacy, but they keep their speciality from fructifying, which would spoil it for their purpose.

Much has been added to our knowledge of the Leaf-cutting Ants by Mr. Beebe, who, in his fascinating book *The Edge of the Jungle* (1921), gives us an account of his own observations of a species of Atta in British Guiana. He had the good fortune to see, at one time, a royal procession leaving the nest in preparation for the nuptial flight. The great queen laboured painfully up to the tunnel far away from the real entrance to the nest. Behind her came the kings, much smaller than she, but large in comparison with the workers that ran all about them. When the queen reached the surface she poised herself on the tips of her slender legs and stretched out her great wings, looking like an aeroplane in miniature. Immediately the little workers swarmed over her, inspecting every organ, cleaning her antennæ,

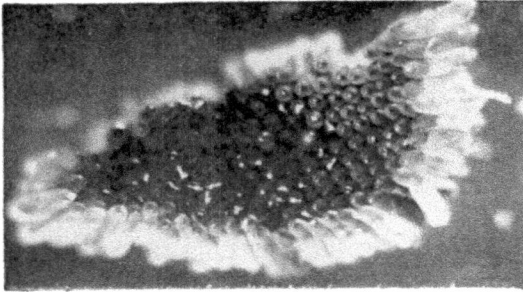

FIG. 1.—The egg-mass of the Mosquito floating on the water. It consists of nearly 300 eggs arranged in the form of a little raft.

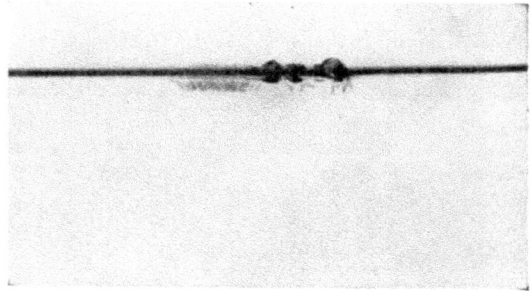

FIG. 2.—Four egg-rafts of the gnat with the young larvæ just emerging into the water.

FIG. 3.—Six hours later nearly a thousand larvæ have emerged from the four egg-rafts.

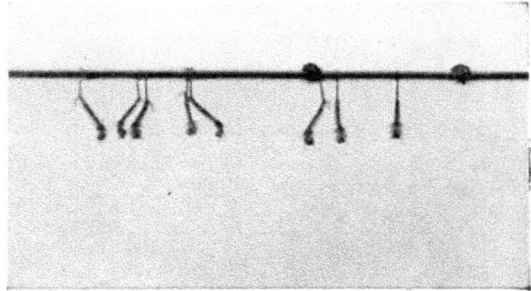

FIG. 4.—The gnat larvæ when four days old. They are resting at the surface and taking in air by their tail-tubes.

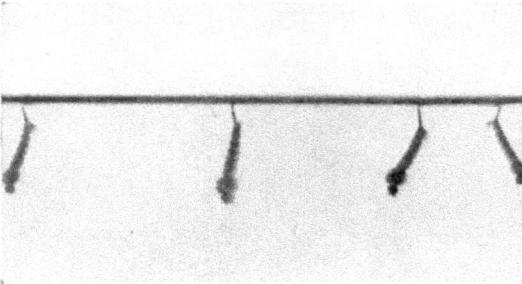

FIG. 5.—When ten days old the larvæ are full-grown. They are still hanging from the under-side of the "surface film" and obtaining air from above by means of their tail-tubes.

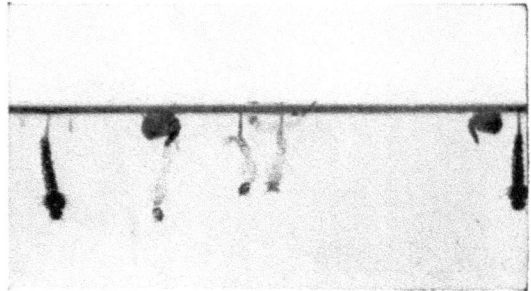

FIG. 6.—On the eleventh day they suddenly moult their skins and change into pupæ, which usually rest passively at the surface, breathing by tubes on the prothorax, but at times show an activity unusual in insect pupæ. One pupa is seen just in the act of casting off the larval skin. (See Fig. 7.)

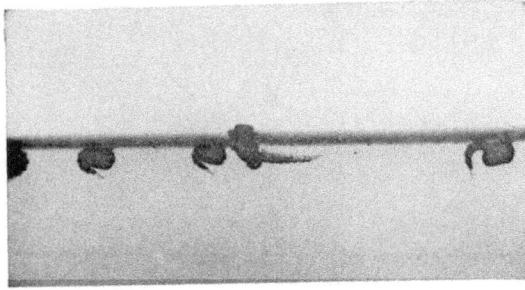

FIG. 7.—On the fourteenth day the pupa bursts its skin at the surface of the water, and the perfect gnat begins to emerge,—

FIG. 8.—and slowly rises into the air, which is henceforth its proper element,—

FIG. 9.—finally withdrawing its legs and wings from the last encumbering folds of the pupa skin.

FIG. 10.—Two minutes later it is trimming its wings in readiness for flight.

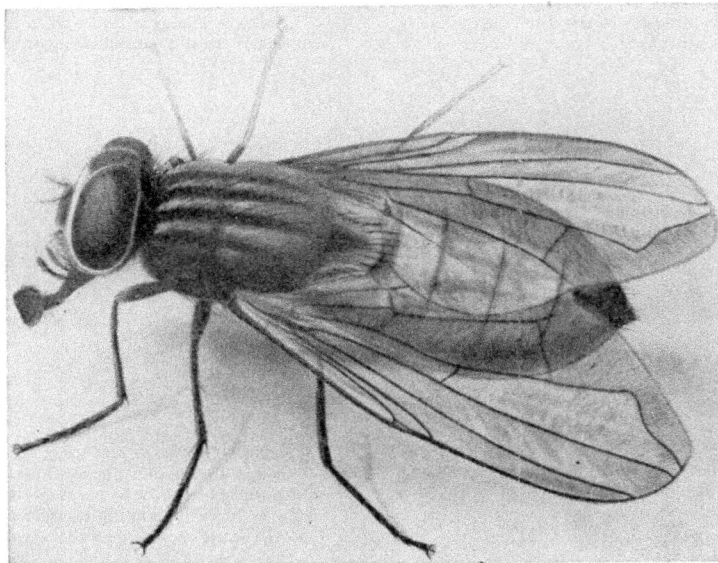

Photo: James's Press Agency.

HOUSE-FLY (*Musca domestica*)

It is bred in rubbish heaps during the summer months. The House-Fly is an agent in carrying typhoid germs, which readily adhere to the numerous hairs seen on the fly's body and appendages. (The Blow-fly, or Blue-bottle, lays its eggs on dead animals, and on joints in the larder if it has access to them.)

legs, and gauzy wings. She endured this for a few minutes, then moved her wings, threw off her load of busy mechanics, slowly rose in the air followed by the males, and was soon lost to view.

But on another occasion the observer was able to follow the story farther, for he saw a queen descend in a long spiral to the ground, rest a few minutes, clean her antennæ, and begin to scrape at the sand with her jaws—the foundation of a new colony at which for many days she labours alone. She plants the little fungus pellet she has carried with her from the old nest in a pouch in the lower part of her mouth, and tends it with the utmost solicitude. The care and feeding in her past life have stored within her the substance for vast numbers of eggs. Nine out of ten she lays, she eats to give her strength to go on with her labours, and when the first larvæ emerge they too are fed with surplus eggs.

There are three castes of workers, large "soldiers," ordinary "workers," and small "workers," or, as Mr. Beebe names them, Maxims, Mediums, and Minims. The first brood, which hatches out in about six weeks, are all minims, and they take charge at once of the fungus, enlarging the nest, attending to the queen and young, and other domestic occupations. When the larger workers emerge, foraging and leaf-cutting begin. In bands they issue forth and search about until they find one of the ant trails trodden down by millions of their kind before them, and stream along it till instinct impels them to climb a tree and drives each ant out upon a leaf.

Standing firmly on the leaf he measures his distance by cutting across the segment of a circle with one of his hind feet as a centre. . . . He does not scissor his way across, but bit by bit sinks the tip of one jaw, hook-like, into the surface and brings the other up to it, slicing through the tissue with surprising ease. . . . Holding his bit of leaf edgewise he bends his head as far down as possible and

secures a strong purchase along the very rim, when as he
raises his head the leaf rises with it, suspended high over
his back out of the way.

From this the ant gets its popular name of Parasol Ant.

Mr. Beebe, with due precautions against attack by the
insects, which are formidable collectively, dug out a large nest.
At first only workers came forth, but by and by the large, one-
eyed, round-headed soldiers lumbered forth to battle, and attacked
his well-greased boots. He tells us that their bulldog-like grip,
which does not relax with death, is taken advantage of by the
Indians, who use them for stitching wounds, applying their jaws
to the apposed edges of skin and then snipping off their bodies.
As we have mentioned, the leaves the ants bring in are not eaten,
but are masticated to a pulp and used as a fertiliser on which to
grow the fungus which is their only food—indoors at least.
Three feet down the great corridors opened out at intervals to
chambers as large as a football, which were filled with the soft
whitish mould which is the *raison d'être* of all the ants' labour.
In one of these chambers Mr. Beebe found groups of workers
in the act of chewing up the leaf-pulp.

The Ways of Army Ants

Of great interest, too, is Mr. Beebe's account of the habits
of the formidable Army Ants. Discovering a nest of these on
the ceiling of an outhouse, the naturalist made for himself an
observation post by placing, at the cost of several fiery stings,
a chair with its legs in tins of a tarry disinfectant. There, within
a foot or two of these myriads of terrible jaws, he spent many
hours watching the home life of the colony.

The whole structure—foundations, walls, and ceilings—was
made of living ants, their legs stretched out to the utmost, their
bodies erect, and their weapons always in a position of readiness for battle. The entrance was guarded by a mat of living

ants, and near the door the edges thickened and met overhead to form a tunnel through which every returning worker had to pass with her booty.

Returning soldiers dropped their load of plunder near the entrance to be dealt with by the worker. They were then immediately surrounded by a group of workers, who put them through a very thorough scraping and cleaning, and they not only submitted with a good grace, but turned over on their backs to facilitate the process.

Spraying with formol disorganised the colony, which broke up in long festoons, and moved away, carrying eggs and larvæ. Next morning it was found that about a third of the ants had remained on the floor in charge of larvæ at the critical stage of passing into the pupal stage. The workers were very busy gnawing wood to dust, and rags to shreds, to provide the light covering which seemed necessary before the larvæ would begin to spin. The following morning the whole horde had disappeared.

Termites or "White Ants" are not related to the true Ants, but their achievements are equally wonderful. They are abundant in many warm countries, notably tropical Africa. They live together in great communities, sharing a many-chambered earthen nest. The hills or termitaries, which they build are often twice a man's height and strong enough to stand upon. In South Africa telegraph posts have to be made of iron to resist their jaws. There is striking division of labour, as with the Black Termite so abundant in Ceylon. When on the march the Black Termites move in great armies, sometimes comprising 300,000 individuals. It has been computed that there are 200 "soldiers" to every 1,000 workers, the number of soldiers guarding a march varying with the danger. The long troop of workers marches between two lines of soldiers. Their tactics are nothing short of extraordinary: there are guides and scouts searching out new lines for foraging. "Very carefully, step by step just like cats, they slink forward, one behind the other, and if the foremost detects anything the least suspicious, he draws nervously back, pulling his 'brave' comrades after him." There are "soldiers" that restore order in the ranks where there is panic; the orders seem to be given through the antennæ or by a quivering of the whole body. Professor Bugnion tells of a war which lasted for three days. The Black Termites often wage a bitter battle with the well-known Tailor-ant, *Œcophylla;* when the latter draw near the termites squirt full in their faces drops of a secretion or fluid which seems to drive the true Ants almost crazy.

§ 4

THE STORY OF BEES

The Beehive

In the Hive Bees (Apis) we have a further illustration of insect communal life. Whatever the nature of the communal life of bees may be, we cannot liken it to that of human society. The one is run on predominantly instinctive lines, the other is predominantly intelligent.

An element of *permanence* distinguishes their communities, for many workers as well as the queen survive the winter. To the industry and food-storing habit of the Hive Bee is probably due their complex social life; the storing has enabled the community to survive unfavourable seasons and become permanent. When spring reawakens the earth and the willow-trees are be-decked with catkins, and gorse and violets and primroses send out a fragrant invitation, the bee world resumes its busy life again. The workers set to work to "spring-clean" the hive and build new combs of hexagonal cells to accommodate the eggs the queen has again begun to lay. Some of the workers sally forth to bring in fresh stores of pollen and honey, while others are nurse workers in charge of the fast-filling nurseries. In early summer the hive is a prosperous and busy city, inhabited by three distinct types of individuals. The head of the community

is the queen, not by reason of her wits, for her daughters far
surpass her in brains and activity, but because she is the mother
bee, who alone can increase or restore the population.

The Queen

One of the most remarkable facts about hive bees is the
apparently psychical dependence of the community on the presence of the queen. If she is removed, the bad news spreads
quickly through the hive and there is a strange disorganisation.
When the bee-keeper replaces her, the good news soon circulates,
and there is harmony once more. According to some authorities,
the queen has a peculiar odour which is reassuring to the workers.
There is no doubt that smell counts for much among bees.

The queen bee is concerned only with egg-laying; the life
of the hive is sustained by the worker bees, which are active, intelligent, but sterile females, with their reproductive systems in
a state of arrested development. Thirdly, there is the drone
section of the community, the males, who take no part in the
work, and forage only for themselves, and then not sufficiently
to satisfy their greed for honey. It has been said that they com-
port themselves in the hive as did Penelope's suitors in the house
of Ulysses: "Indelicate and wasteful, sleek and corpulent, fully
content with their idle existence as honorary lovers, they feast
and carouse, throng the alleys, obstruct the passages, and hinder
the work." But this is not quite accurate. Drones spend much
of their time flying about very energetically in the vicinity of
the hive. They are on the look-out for an emerging queen, and
they are usually disappointed.

The Bees' Diligence

The stronger workers have to provide food for the whole
colony. Their diligence is immense; they toil from morning to
night with ceaseless energy, gathering in the precious store of
honey and pollen, and it is said that in summer-time the average

life of a worker bee is only about two months. Their brains become hopelessly fatigued. In a colony of 50,000 bees it has been estimated that there are 30,000 workers, and if each makes ten trips a day 300,000 flowers would be visited. About 37,000 loads of nectar are required for the production of a pound of honey.

To obtain the nectar, the bee protrudes its tongue into the flower tube and sucks up the nectar into its mouth and thence into the "honey-bag," where it changes into honey, which is deposited in storing cells for the indoor workers to draw on for themselves and also, of course, for the nutrition of the larvæ. The golden pollen is kneaded into a little ball and carried back to the hive in the "pollen-basket," a little cavity in the bee's hind-leg.

There is a popular idea that bees fly about from flower to flower in a haphazard way, sipping nectar from any blossom that takes their fancy. But as a matter of fact, and as Aristotle noticed, many bees keep as a rule to a single species of flower for collecting pollen and nectar. This is an advantage to both flower and insect. If the bee were to go from one type of flower to quite a different one, time would be lost in locating the nectar. Moreover, when the bee is constant for a while to the same kind of flower-cups, pollination is effected and waste of pollen is prevented. The mutual aid which is an undoubted fact in the bee society sometimes takes the form of showing each other valuable sources of nectar.

The Nurseries

Within the hives the younger workers are busily looking after the nurseries and attending on the queen. The newly hatched grubs are fed on a kind of pap regurgitated by their nurses, but soon they are ready for a more substantial diet of pollen and honey. Then the larvæ spin cocoons and the workers shut the cells with little caps of porous wax, and leave their

Photo: James's Press Agency.

QUEEN OF THE TERMITES AND HER COURT

The queen Termite is about the length of our middle finger. Most of the length goes to the posterior body, which is bloated with eggs. The relatively small head and thorax are seen in front. The queen is lying in the royal chamber of the termitary, the door of which, though originally allowing her entrance, is much too small to allow her exit now. She often lays sixty eggs per minute. She is seen surrounded by a body-guard of workers, and outside these a circle of soldiers.

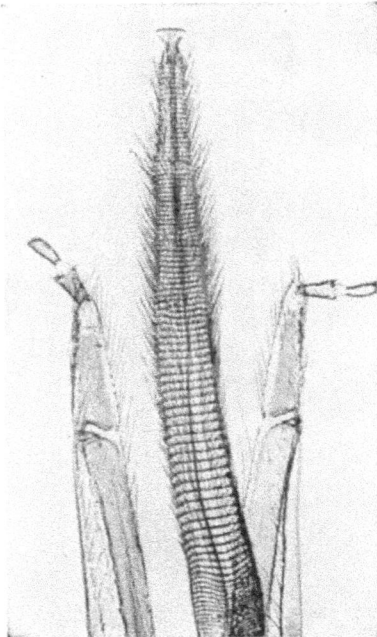

Photo reproduced by courtesy of F. Davidson & Co.

TONGUE OF HIVE BEE

By means of this long tongue the worker-bee is able to extract from flowers the nectar which is afterwards stored as honey in the hive.

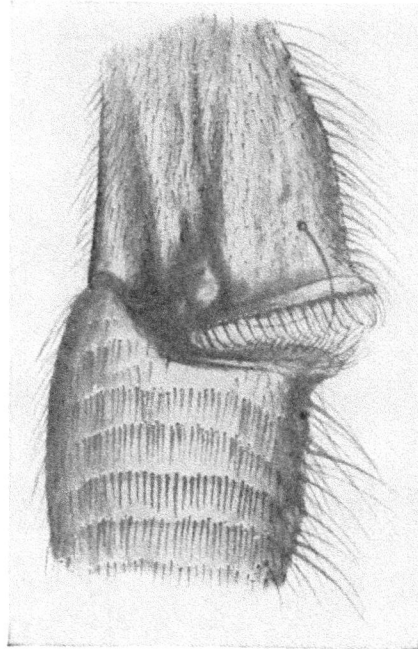

Photo: James's Press Agency.

POLLEN BASKET ON HIND-LEG OF HONEY BEE (WORKER)

Pollen, kneaded into a little ball, is carried from the flower to the hive in receptacle shown.

Photo: J. J. Ward.

TREE WASP

A queen wasp biting off bits of wood which she chews into a pulp—the raw material which she manufactures into the familiar grey "paper" that forms the walls of her nest.

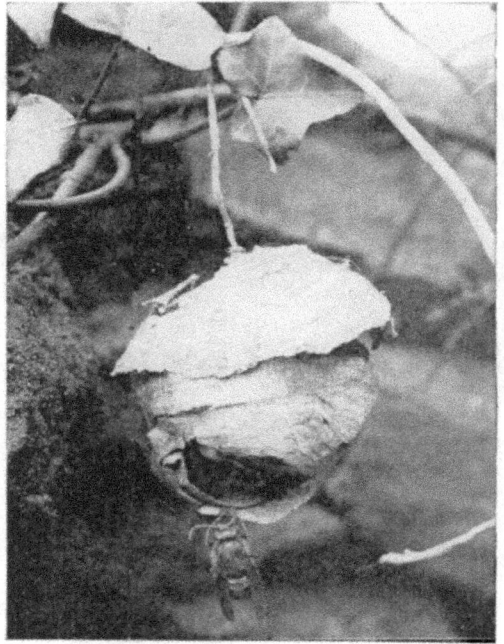

Photo: J. J. Ward.

NEST OF WASP MADE ENTIRELY BY THE QUEEN

The nest was built in between the stones of a wall, and one stone has been removed to expose it. This was the nursery for the first generation of worker wasps, which, when mature, would enlarge the home for the rapidly increasing family.

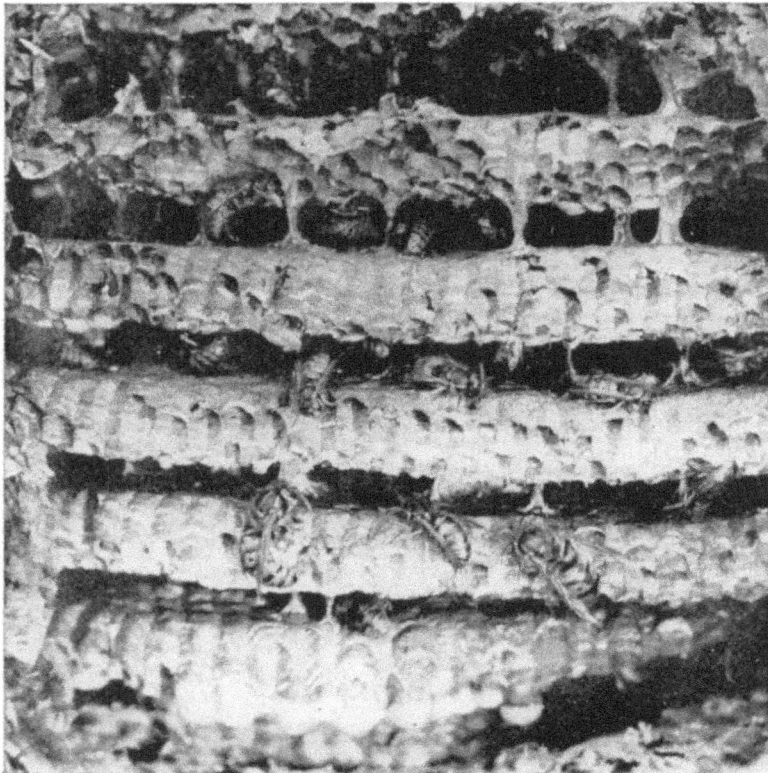

Photo: J. J. Ward.

INTERIOR OF A NEST OF ACTIVE WASPS

On the bottom comb the dome-shaped queen-cells are seen, and on the comb immediately above are two young queens just emerged from the cells, with a male wasp resting on the comb between them. All above are worker wasps, tending the young and making new combs.

charges to a thirteen-day pupation, after which yet another generation of worker bees bite off the roofs of their cradles and join in the busy life of the hive. In larger cells the queen deposits eggs which are not fertilised, and these develop into drones. Still later in the season "royal" cells are constructed, in which the queen lays fertilised eggs, identical with those laid in the ordinary worker cells, but the grubs which hatch out receive a special "royal jelly" from the mouths of their attendants, instead of the usual fare of masticated pollen, and the effect of this diet is to make the grubs develop into "princesses" instead of workers.

It should be noted that a queen bee receives from a drone in the course of her "nuptial flight" a store of sperm-cells with which she may fertilise the eggs she lays during the next year or more. It depends on the egg-laying movements of the queen whether the laid egg is fertilised or not.

The Swarm

Then comes the remarkable upheaval of the busy hive—the departure of a "swarm" headed by the queen bee. Whether swarming is due to the overcrowded state of the hive, or to the queen's excitement when her young rivals are stirring in the royal cradles, or to a sudden desire on the part of the workers, a harking back to the time when there were no hives and motherhood was not given only to one among thousands, a desire to break out of their "prison bounds of order, commendable toil, chill, maidenly propriety," who shall say? But suddenly the routine of the hive is broken through, work is suspended and many of the workers become restless and excited, and gorge themselves with honey till at a given signal the swarm issues from the hive, "in a tense, direct, vibrating, uninterrupted stream that at once dissolves and melts into space, where the myriad transparent furious wings weave a tissue throbbing with sound."

The mad, joyous dance in the sunlight over, the swarm

returns to earth, and now there is the morrow to consider and a new home has to be built. Scouts go out, and when they have found a suitable site the workers at once begin to fashion a new comb, in which the queen lays eggs, and so a new city springs up. The hexagonal cells of the comb are made of thin plates of pliable wax, which comes from little pockets on the bee's abdomen. To start the secretion of the wax great heat is needed, so the bees gather together in a great pendant mass till "a strange sweat, white as snow and airier than the down of a wing, is beginning to break over the swarm." The worker bee removes the wax scales from her body with a pair of pincers she has at one of her knee joints, and then chews them into a soft paste which can be moulded into the delicate fabric of the cells.

Honeycomb

The bees' comb is one of the wonders of the world. In spite of its extraordinary fragility it is able to suspend a weight thirty times as great as its own. A small block of wax attached to the roof of the hive makes the foundation, from which the layers of cells grow out downwards and sideways, leaving a gangway for the streams of bees to pass to and fro. The usual shape of the cells is hexagonal, individually well suited for the cylindrical body of a grub, together ideally constructed to prevent waste of space. But bees adapt themselves to unusual circumstances and build triangular, square, or other cells in odd corners if the need arises. The cells are not quite horizontally placed, having a slight upward tilt which prevents the spilling of thin honey. Extreme delicacy of touch is required in the moulding of the plastic wax, for the 1-180 part of an inch is the thickness of the tissue-paper-like cell-walls.

The Nuptial Flight

While the new colony is rapidly growing up, life continues in the old hive; it is, in fact, about to renew its youth. One of the princesses is awakening, and the remaining workers are watching over her. She appears from the shelter of the royal nursery, and the workers brush her and clean her and caress her. Impelled by some strange instinct, she immediately seeks the other cradles, tears open the cells and relentlessly stings her sisters, her possible rivals, to death! A few days later, on a bright and sunny day, she leaves the hive for her nuptial flight. She soars aloft into the blue sky followed by a crowd of drones from neighbouring hives, and somewhere in the solitude of the blue the strongest male overtakes her and meets love and death in the same instant; and the bride-widow returns to the hive.

Massacre of the Males

For the remainder of the summer the busy life of the hive goes on as before, the queen perpetually egg-laying, the workers foraging and nursing, the drones leading a life of ease. But one day the decree goes forth that those that do not work shall not eat, indeed shall not live; and the massacre of the males begins. Vigorously and pitilessly the long-suffering workers at last turn on the drones and slay them all.

Flowers are becoming scarce, and the days are short and chilly, so the bees cease their labours and prepare for the long sleep of winter, if sleep it can be called, for the life of the hive is slackened, not completely arrested. The bees gather together in a great cluster, with their queen in their midst, and by the beating of their wings they keep up a current of warm air. The bees nearest the store cupboards pass the honey to their neighbours, and so food is circulated through the drowsy mass, enough to keep the fire of life glowing, ready to burst into flame again with the return of spring.

Among different kinds of bees there are different degrees of sociability. Some, such as the Leaf-cutting Bee, are quite solitary; others show a certain amount of co-operation combined with a large amount of independence.

The Humble-Bee

The Humble-Bees (Bombus) live in communities which last for one season only. The queen humble-bee, after her autumn nuptial flight, creeps into a hole under a sun-warmed bank and there lies torpid thoughout the cold weather. Spring awakens her and she sets to work to prepare for her expected brood. She secretes wax, makes a few cells, and lays her eggs in these. She has herself to discharge the whole labour of foraging for honey and pollen, keeping the cells clean, kneading the bee-bread, and feeding and tending the hungry larvæ. She is a queen in the sense of being the mother of the whole colony, but she is a very hard-working queen for a time. Later, when the first batch of young ones, which are always workers, are fully developed, they take the domestic details on themselves, and the queen can now devote herself to her true business of motherhood. As in the case of wasps, the community dissolves at the end of the summer, workers and drones all dying, but a few young queens surviving through the winter to found the colonies of the following year. In this and in many similar cases it is difficult to know whether one should speak of a large family or of an incipient society.

§ 5

The Story of Wasps' Nests

Even Solitary Wasps instinctively provide for their young, though they die before these hatch out. They deposit the eggs in a shelter and leave with them a larder of fresh meat, in the shape of living insects rendered unresisting by the paralysing effect of the wasp's sting on their nerve-centres. The Social Wasps live in communities which last from spring to autumn. Winter is the time of inactivity, but in some secluded spot, a cranny in a wall or a sheltered nook in a rubbish heap, the queen wasp, who mated at the end of last season, is sleeping her winter sleep, tiding over the hard months in a state of passiveness in much the same attitude that her body assumed during the pupa stage. With the coming of spring she reawakens, and the season's activities are soon in full swing. The queen's first care is to choose a suitable site for the nest she is about to build, and a cavity in the shelter of the gnarled roots of an overthrown tree is as good as any. Then she sets to work to collect wood-fibre, which she rasps with her jaws from posts and palings. This wood-pulp she kneads with her saliva into the "paper" with which the nest is built. She spreads the first layer on the root she has chosen as the foundation from which to hang the structure, and gradually, hour by hour, pellet by pellet, she moulds a disc, and then a stalk, and then a canopy to shelter the first layer of cells. In each cell as it is completed she deposits an egg, which she cements to the cell-wall, for the open end of the cell is directed downwards.

In a few days the legless grubs emerge, and the queen becomes a nurse as well as a home-builder, until the older grubs mature, and a staff of worker wasps is ready to take on the manual labour and allow the queen to devote herself to egg-laying. The workers add to the original comb and suspend a new storey from it by little stalks. One storey is added after another. The rounded outer covering is also extended, by being hollowed out inside and added to outside. This outer envelope may consist of as many as a dozen layers of the paper, which is a water-proof and non-conducting material, so that the necessary temperature for the development of the young is kept up. The entrance opening of the envelope is always at the foot of the pendant nest, and all the openings of the combs point towards it, so that the young are reared in inverted cradles.

The young wasp grub at first keeps its position by clinging with its tail to the egg envelope while it pokes its head out for food, but later it uses its jaws and a sort of sucker-foot on its tail as grasping organs. If it does happen to fall out, the worker nurses will probably throw it out of the nest, just as they do with rubbish when they are cleaning. The first thing the fully formed young wasp does, if it has safely passed through its head-downwards larval and pupal stages, is to crawl about and visit the grubs, tapping them on the head till they emit a tiny drop of fluid, which the young wasp licks greedily. Then it is ready to help its mother with the housework, and in a few days is strong enough to go out on foraging expeditions. The mother wasp also visits the grubs for this delectable drop.

How the Wasp Works and Dies

The young wasp's duties at first consist mainly of paper-making and building, for the nest is continually growing. She works backwards so that she does not tread on the newly applied pulp, and she moulds her material to the proper thickness, testing it with her feelers. But after a week or two her salivary glands are exhausted, so that she has to give up the manufacture of paper and turn to the older wasp's task of caring for the young, feeding them with the soft parts of insects and occasional sips of fruit-juice or nectar, and cleaning them with care. So through the summer the busy life of the community goes on. The queen has laid thousands of eggs, and a great army of her daughters is engaged in enlarging the nest—which may now have seven or eight tiers or combs enclosed in a great ball of grey paper—in keeping it scrupulously clean, and in caring for the rising generations. Some of these workers, though they are never impregnated, may occasionally lay eggs, which, like the unfertilised eggs of the queen, invariably develop into males.

As summer wanes, the workers build larger cells in the lower combs. These are the royal nurseries in which a brood of perfect females, not sterile workers, and males are reared. On this brood the future of the race depends. A few weeks, and a great change takes place—summer is still here and the wasp colony is at the height of its prosperity, a healthy, active community; then the chill finger of autumn passes over it, and the first shiver marks the beginning of the decline of the colony.

Prosperity is succeeded by starvation. There are no stores to fall back on, and deadly numbness and demoralisation break down the orderly routine of the nest. The exhausted workers die in their thousands, and with them the parent queen. None but the young royalties survive, and the males only long enough to mate with the young queens; thereafter they also die. The young queens destined to found new colonies next spring alone escape the common fate, but the demoralisation shows itself in them too, for they devour the remaining eggs and larvæ, and on this rather cannibal fare they are able to survive the winter.

§ 6

LIFE HISTORIES

Story of Cabbage White Butterfly

The food of Insects is extremely varied, not only in different species, but also within a single life-history, and it naturally follows that there is much variety in the ways of obtaining it, and, in particular, in the structure of the appendages associated with the mouth. Insects depend greatly on their sense of smell when in search of suitable food, and the organs of smell, minute olfactory pits or bristles, are found chiefly on the antennæ. Some insects move their feelers markedly on coming near strong-smelling substances, and some are unable to find their appropriate food without the aid of their antennæ. For instance, Carrion Beetles which had had their antennæ removed were found to be incapable of locating their evil-smelling food. A very striking example of change of diet is seen in the life-history of a butterfly, such as the common Cabbage White Butterfly. The small, sculptured eggs are laid in large numbers on the plant which is to form the food of the caterpillars. The caterpillar emerges from the egg as a worm-like, short-legged little animal, green against the green of its natural haunt, with simple eyes, short feelers, stumpy abdominal "pro-legs" in addition to the three pairs of jointed

thoracic appendages, and strong, hard jaws well suited for gnawing green food. Its business in life is to feed and to grow, and
it feeds rapidly and almost continuously. It may eat many
times its own weight in a day, but probably only digests the fluid
part of the food. It outgrows its inexpansible chitinous covering, and has to moult it, an exhausting and dangerous process.
Then it feeds and grows and moults again, until at its limit of
growth it passes into a resting phase. It becomes a pupa, or
chrysalis.

The Cabbage White Butterfly larva suspends itself in a
quiet corner by a silken thread, with its tail against a support,
and the larval skin forms the pupa case, but many other pupæ
(e.g. many moths) have the additional protection of a cocoon,
either of pure silk secreted at the jaws, or of silk mixed with
leaves, moss, or other extrinsic matter. The larva (i.e. the caterpillar) now undergoes the great change which is called metamorphosis. Within the cocoon the body of the larva is broken
down and is built up again on a new architectural plan. When
the reconstruction is completed the fully-formed insect emerges.
What a contrast! It is now an intensely active butterfly, having
left behind it the shrivelled skin of the creeping caterpillar, and
for a brief season it lives its aerial life, growing not at all, feeding but little, and then only on liquid nectar by means of the
long sucking-tube so different from the strong biting jaws of the
caterpillar: hunger is no longer the preoccupation; the butterfly lives for love, and before it dies it deposits its eggs on the
green plant which it cannot itself eat, but which forms the right
food material for the offspring it does not survive to see.

Beetles

Beetles are essentially biters, with very strong and hard
mouth-parts, one pair of which, the mandibles, are sometimes
of relatively enormous size, with sharp saw-like edges. Many
of them, such as the Weevils, are vegetarians, feeding on green

AN ILLUSTRATION OF THE GARDEN OF GETHSEMANE BEFORE THE COMING OF THE LOCUSTS

THE GARDEN OF GETHSEMANE AFTER LOCUSTS HAD PASSED.   EVERY FLOWER AND EVERY SHRUB WAS LEFT
QUITE BARE

AN ASIATIC LOCUST

After several moults the locust is a perfect, winged insect. Immediately after the skin-casting it is soft and helpless, but a short time in the sunlight makes it firm and vigorous. The lower wings are fan-shaped, and fold up longitudinally under the longer, narrow and spotted outer wings.

plants or on the bark and wood of trees, but many others are carnivorous and destroy numbers of wireworms, "leather-jackets" (the larvæ of the "Daddy-long-legs"), Saw-Fly larvæ, and other insects which are detrimental to crops. Others, again, feed on the decaying flesh of dead animals, and the busy "Burying Beetles," which join forces in their work, act as useful bands of scavengers.

An Important Linkage

Other groups of Insects, with quite different mouth appendages, belong to the sucking types, which feed on liquid food. Instead of cutting, toothed jaws, they have sucking-tubes, often accompanied by sharp piercing needles as in the Mosquito, which pierce the skin and suck in the blood of the victim. The nectar of flowers is another great source of liquid food, and is sought by Bees, Butterflies, Moths, and others which have sucking mouth-organs. Perhaps the most important linkage in the whole system of animate nature is the linkage between flowers and their welcome insect visitors. For these visitors secure cross fertilisation, and this is often essential to seed-bearing.

§ 7

Life-histories

There are various ways in which the young forms of Insects hatch out from the shells within which they develop. Some caterpillars eat through the shell; some maggots wriggle until it breaks; and some larvæ have special instruments for the purpose. Thus the larval flea has a temporary piercing organ on its head. Many larvæ differ markedly from the adult forms, and they are of several different types; they may be active, long-legged, flat-bodied (campodeiform) larvæ (very like the primitive Bristle-tails), e.g. many beetle larvæ, May-Flies, Stone-Flies, etc.; or they may belong to the more wormlike (eruciform) group, such as caterpillars (i.e. young of moths and butterflies), maggots, and various grubs, and these may be more sedentary in habit. In the course of the life-history of many insects a marked change of form takes place—metamorphosis. According to the degree of metamorphosis, insects are divided into three groups: (1) When *no* metamorphosis occurs and the young are hatched as miniatures of the adults, e.g. the most lowly insects, the Springtails and Bristle-tails. (2) An intermediate group comprises those insects which show partial metamorphosis. In this type the insect is able to move and feed practically throughout its development; the change is a *gradual* one. Through a series of moults, made necessary by the inexpansible armour of chitin, the insect reaches the adult condition.

For instance, the young Locust, as it emerges from the egg, has a pale, soft body swathed in transparent skin. It sheds its mantle, and, gaining strength in the sunlight, becomes firm and black, only differing from its parents in size, colour-markings, and the absence of wings. It feeds hungrily on vegetable substances, and grows and moults, each moult leaving it larger, brighter, and hungrier than before, until after the third moult its wings begin to show. The moulting process lasts only about half-an-hour, and the locust only stops feeding for a few hours. No phase of torpor or quiescence occurs in this "half-metamorphosis" type, and after the fifth skin-casting the locust is a perfect, winged insect, soft and helpless and very vulnerable for a time, but rapidly regaining firmness and vigour.

(3) When complete metamorphosis occurs, a quiescent pupal or chrysalis stage comes between the larval and adult stages. Growth occurs during the larval stage, a period of voracious feeding, rapid growth, and numerous moults. The larva eats far more than is necessary to maintain its life, and lays up a reserve store which provides for the resting pupal stage which follows. The pupal stage is a time of little or no external activity but great internal changes. The larval tissues are broken down and their substance is reconstructed into the very different

tissues of the adult. From the pupa case the adult insect emerges, different in form and habit, winged and aerial. Metamorphosis implies far more than the acquisition of wings, and one of the most marked differences between larva and adult is in most cases the difference in the food and the method of taking it. This is so great that the transition from larval to adult habits could not take place along with continuous external activity—the quiescent period of reconstruction is essential.

§ 8

Insects and Man

A great many insects live their busy days and perish without affecting man at all, except that they delight him with their exquisite colours and markings and interest him with their ways, but some are his friends, and perhaps more he reckons his foes. Even the Bee he too often shrinks from, remembering the weapon she carries and forgetting her honey and the infinite service she renders by securing the pollination of many flowers. The Termite may be as much a tiller of the soil as the Earthworm is, but she attacks his furniture and the wood of his house; the cochineal and "lac" that insects provide are relatively insignificant; and "locusts and honey" may be thought a dainty dish in the East, but a locust swarm will blight every green thing in a district. "He scatters the seed, and when he looks for green heads to appear, the earth opens, and, lo, an army of long-faced, yellow grasshoppers come forth!"

Locusts

Wherever locusts are resident they do a great deal of damage, but it is their sudden migratory swarms which are so disastrous. They increase in numbers during favourable seasons; then, one year, when the food-supply is insufficient, they collect in immense swarms and travel long distances, devouring every green thing in their path. A tobacco-grower saw a swarm of locusts descend on a plantation of forty thousand young plants. Twenty seconds later not a leaf remained! The Old Testament speaks of the locust as one of the plagues of Egypt. "They covered the face of the whole earth, so that the land was darkened; and they did eat every herb of the land, and all the fruit of the trees which the hail had left: and there remained not any green thing in the trees, or in the herbs of the field, through all the land of Egypt."

In addition to the formidable list of insects, larvæ and adults, injurious to plants, another list must be added of those which affect the health of man and of his stock. There are a number of ways in which insects may affect the health of man. They may have poisonous bites or stings, as in the case of certain Bugs, Bees, Wasps, etc., which cause inflammation and sometimes feverishness; or they may be parasitic, either true parasites such as fleas and lice, or accidental parasites, such as fly-maggots, which sometimes reach the stomach and cause great pain; again, they may carry disease germs. Most important of all are the cases in which an insect is an essential host in the development of a disease-producing organism, without which the life-history of the organism cannot be completed.

For example, the Mosquito is not only the means of introducing into the blood of man the Protozoön which causes malaria, but the life-history of the malaria organism cannot proceed without the insect; the different stages can only be reached within the bodies of man and mosquito respectively, so that the extermination of mosquitoes would wipe out malarial fever. In other cases, the insect is not necessary to the life of the disease-producer, but acts as a transmitter, as in the case of plague, where the bacillus is conveyed from rats to man by means of ratfleas, which inoculate the victims while biting. Further cases of disease-carrying form another list—those of the simple carriers, such as the common House-Fly: it is not a blood-sucking insect, but it has a body and legs thickly covered with hairs particularly well suited for transferring germs, such as those of typhoid fever, from place to place, and it thus brings the microbes of the garbage heap to its next feeding-place, our dinner-tables. There is a long list of diseases in which insects play an important part—typhus fever and Lice, sleeping sickness and Tsetse Flies, relapsing fever and Lice, and many others. Many insects also affect the domestic animals, for example the "bot-flies" which cause severe boils and other disorders in cattle.

Such examples out of the list serve to show some of the complex inter-relations between man and Insects, and to indicate some of the aspects of the struggle for existence. Man's enemies are innumerable; he tames the wild beasts, and domestication brings its own penalty, for a sucking insect wipes out a whole herd; he exterminates great flesh-eating animals that would rival him, but a common house-fly brings microscopic germs to his table and spreads death through his cities. It is hardly too much to say that the tendency of injurious insects to prolific multiplication is a continual menace to civilisation, and this should lead us to attach increasing importance to the preservation of the numerous insectivorous birds which maintain the balance of Nature. But this subject will be discussed in a special article dealing with Inter-relations.

BIBLIOGRAPHY

BALLARD, *Among the Moths and Butterflies.*
BASTIN, *Insects: their Life-histories and Habits.*
CARPENTER, *Insects: their Structure and Life.*
EDWARDES, TICKNER, *The Lore of the Honey-Bees.*
FABRE, *Insect Life, The Life of the Fly, The Life and Love of the Insect,* etc.
LATTER, *The Natural History of some Common Animals.*
LUBBOCK, *Ants, Bees, and Wasps.*
LUTZ, *The Field Book of Insects.*
MAETERLINCK, *The Life of the Bee.*
MIALL, *Injurious and Useful Insects* and *Life-history of Aquatic Insects.*
SHARP, *Cambridge Natural History* (two volumes on Insects).
SLADEN, *The Humble-Bee.*
WHITE, *Ants and their Ways.*

XV
THE SCIENCE OF THE MIND

THE SCIENCE OF THE MIND

THE NEW PSYCHOLOGY—PSYCHO-ANALYSIS

I T is something of a paradox that the most difficult thing the mind finds to master is the mind itself. In recent years science has applied itself to the problem with a new keenness; much attention has been given to the special study of the mind of the child, and valuable results have been obtained from the study of animal behaviour. In particular there have been many investigators at work on what has become known as the New Psychology, which concerns itself largely with abnormal mental phenomena and subconscious operations—that part of mental activity which lies beyond the region of normal consciousness.

Practically all the recent work in psychology has gone to show that there are elements in our minds of which we are unconscious, and that these elements often take a greater share in shaping our behaviour than do the elements of which we are directly aware. The conception of the human mind has, in fact, undergone a profound change; it is revealed as a larger and more complicated affair than we had supposed, and we now see that what we had taken to be the mind, is, in reality, a superficial although very valuable part of man's total mind.

The Senses

Sense-experience forms the foundation of our mental life. In the course of long ages of evolution our sense-organs have evolved, and have given rise to that wonderful organ the human brain. It is through the senses that all the materials with which the mind builds up the higher forms of experience—memory, imagination, and thought—are obtained. For the senses are the Gateways of Knowledge.

It would be going beyond the scope of our subject to describe fully the evolution of our various organs of sense—the mechanism of the eye, the ear, and so on. By these instruments we are able to image and focus the world outside of us.

A sensation depends on some physical influence, the stimulus affecting some part of the outer or inner surfaces, or tissues of the body. In most cases there is a special organ adapted to receive the stimulus, and so to transform its action into a nerve-impulse for transmission to the brain, such as the eye, the ear, parts of the skin, and so on.

The acquisition of the sense of sight vastly enlarged the horizon and widened the mental range; and so with hearing, which is the most recently acquired of our specialised senses. We know that the senses are not infallible; they are limited and imperfect; but there is no evidence whatever that the development of our senses has reached finality.

The Brain

The structure of the brain was briefly referred to in the section dealing with physiology. We need recall only that there are several main divisions of the brain, each with its own peculiar functions. The brain proper consists of the cerebrum, or larger brain, which occupies the whole of the upper and front parts of the cavity of the skull. It is divided into two great cerebral hemispheres, right and left, which are linked together by numerous nerve-fibres. The outer surface or cortex of the fore-brain is the seat of sensation and volition. It is a wrinkled or convoluted fold of grey cellular matter, which if smoothed out would cover a little over a foot and a half square. There are in the convoluted part of our fore-brain (the cerebral cortex) five

THE THINKER

From a Statue by Rodin.

or six times as many nerve-cells as there are human beings in the world, and the complexity of inter-relations is past all telling.

The cerebellum, or lesser brain, lies at the back of the head, and below it is the medulla, whose functions have been previously explained. We need, not, therefore, further enlarge on the outline of our nervous system—the cerebrum, cerebellum, brainstem, spinal cord, and nerves. "That marvellous structure, the human brain, is the product of millions of years; its history begins with life itself." The brain is a republic of nerve centres; each part has its own peculiar function—and all in inter-action. There are parts of the brain whose function is unknown—parts which we believe serve for memory, judgment, and imagination. There is reason to suppose that one part is the seat of the processes associated with remembrance of articulation; that another is similarly associated with memory of the sound of words; yet another part of the brain is associated with visual images of words and letters.

There is no lobe in the brain that is the seat of intelligence. It is the whole cortex, we might almost say the whole nervous system, or the whole body, that is concerned in intelligence, not any single region of it. It is by the plasticity, the power of adapting itself to new ways of learning, registering, and repeating new co-ordinations of actions, that the brain is marked out from the rest of the body and even from the rest of the nervous system. Great ability, great intelligence even, are not dependent primarily on the brain.

§ 1

Mind in Evolution

When we look back over the vaguely discerned evolution of Animal Behaviour, we find that it had its starting-point in the tentative movements of simple creatures, as has been explained in a previous chapter. We see such tentative movements in the very lowliest creatures (see Vol. I., p. 76).

At an early stage there must have been established a number of particular answers (involuntary muscular and nervous movements) to stimuli, which became enregistered in the creature, and these ingrained capacities increase in number. We discern a persisting state of the organism which *varies* the answers; there is probably a simple expression of conation or endeavour. And in time we come to perceive something of purposive behaviour. "With the establishment of a nervous system there was opened up the possibility of a new kind of organisation—that of reflex actions and tropisms, which play an important rôle in behaviour, an organisation which heredity perpetuates." Reflex actions are automatic movements of nerve-cells and muscle-cells of lower animals, which secure a fit and proper answer to a recurrent stimulus. Tropisms are on a somewhat higher plane; they are forced or obligatory movements of the animal *as a whole,* that is to say, every creature of the same kind, and in the same physiological state, will behave in the same way. On a still higher level we have instinctive behaviour,

> which reaches its purest expression in ants, bees, and wasps. In birds and mammals it is more likely to occur in co-operation with intelligence. Instinctive behaviour agrees with reflex acts in not requiring to be learned, in being dependent on hereditary nervous predispositions, and in being exhibited approximately in the same way by all similar individuals of the species.[1]

We have discussed previously the history of these progressive evolutionary advances, culminating in intelligent behaviour, and we saw wherein lay their survival value. We need not consider them further here. Reflex actions, tropisms, and instinctive behaviour have become part of the inborn *hereditary* constitution of all higher animals.

The question may be asked, what, besides what we call our mental faculties and our instincts, forms part of our natural in-

[1] J. Arthur Thomson, *The System of Animate Nature.*

heritance; in other words, what comprises the innate constitution of the human mind? The question is not easy to answer. Dr. McDougall puts the question in the following form:

> Does the native basis of mind comprise any disposi-
> tion, in addition to those which enter into the composition of
> the instincts; and if so, to what extent are they systematically
> linked together?

> We cannot answer this question with a negative. There
> is certainly much beside the faculties and the instincts com-
> prised within the native basis of each human mind. If there
> were not, it would be impossible adequately to account for
> the vast superiority of mind of the human adult to that of
> the highest of the animal. Some of those who regard the
> mind purely from the physiological standpoint, and who be-
> lieve that all we have called the structure of the mind can be
> adequately described in terms of the organised structure of
> the brain, take the view that the superiority of the native en-
> dowment of man consists, chiefly or wholly, in the presence
> in the brain of the infant of a great mass of unorganised
> nervous tissue which offers unlimited possibilities of pro-
> gressive organisation. But, even if we accepted the assump-
> tion that the structure of the mind can be wholly described
> in terms of nervous disposition and their connections, we
> could not accept the view that nothing of the mental organ-
> isation beyond the instincts is innate.

The bearing which all this has on our present problem is this: can we say that the particular kind of activity known to us as thinking, feeling, and willing is implicit in the germ-cell just beginning to develop into an organism of great complexity —an individuality in the one-cell phase of its being, a mind-body or body-mind telescoped down. It varies, it makes experiments, it makes its own essays (in internal rearrangement) in self-expression.

> The germ-cell is a sort of a blind artist; its sketches are
> submitted to the criticism of the fully formed organism, the

seeing artist, who will put them in the proper light and bring out what there is in them of value.

If the amœba has in its small way a mind, an aspect of itself corresponding to our mind, and if the amœba uses it when it goes hunting—two not unreasonable hypotheses— then it may be that the germ-cell has also its analogue of mind—a not unreasonable hypothesis, since it develops into a creature with a mind.[1]

§ 2

It is not the province of psychology to explain what mind is; that belongs to the region of philosophy. Still, the great problem which holds an interest for us is concerned with the relation that exists between mind and body. Is mind independent and distinct from the body, or is it merely "an activity of the brain-cells, a product of nerve stimulation"?

Mind and Matter

Men have argued endlessly on the relation of mind and matter. To discuss, even briefly, the various theories—and there are many—would take a volume.

What the precise connection between mind and body is, no one, as yet, has been able to say with any degree of certainty. On the *mechanistic* view, as it may be called, the mind is a direct product of the brain, and has no separate independent existence. Every act of intelligence, every mental activity, is due to a physiological mechanism. Every thought

> is the result of chemical or mechanical changes in the brain; an "idea" is but an explosion or discharge of the brain-cell, an emotion is an activity of the brain bursting into flame; every feeling of love, aspiration, or fear can be explained as due to purely physical changes which produce the vapour of thought, or the aroma of virtue.

If it be held that during life "all mental processes have their

[1] J. Arthur Thomson, *The System of Animate Nature.*

ATTENTION

From the painting by Sir John Everett Millais entitled "The Boyhood of Raleigh." The future navigator is seen listening to a tale of the sea. The intensity of attention is well illustrated. In the act of attention all conflicting tendencies are driven from the mind.

Photo: Rischgitz.

ARISTOTLE

The most profound thinker of the Ancient World.   He has been described as the best educated man of any age.

PROFESSOR FREUD

The "New Psychology" is closely associated with his name.   His doctrines are explained in the section dealing with Psycho-analysis.

Photo: L. E. A.

THE LATE PROF. WILLIAM JAMES

Professor of Psychology at Harvard, and investigator of psychic phenomena in America.

Photo: Elliott & Fry, Ltd.

DR. WILLIAM MCDOUGALL

Professor of Psychology, Harvard University, author of several important works on psychology, including *Body and Mind, a History* and *a Defence of Animism.*

physiological concomitants, it is clear that these physiological concomitants, namely the molecular changes in the nerve-centre, would, if completely ascertained, afford an accurate index of the mental processes."

But no one has ever shown what the chemical or mechanical changes are by which thought and feeling are produced. Mechanism, as applied to mind, remains a mere hypothesis, an hypothesis, it may be added, to which philosophy gives no support.

Another view is that mind is a *separate existence.* The relation of the mind to body is, on this view, frequently held to be one of parallelism: the two series, mental and physical, are independent of each other, each runs its own course, "as two railway trains running side by side on a double track, or two rays of light projected towards the same infinitely distant point, run parallel with one another in time and space." There is no cross effect from one to the other; each is a closed system, with its own laws. When consistently held, this view does not carry us much farther than the first view; each point in the mental series must have its counterpart in the physical series; the laws that are established for the physical must also account for the psychical events.

A third view is *Animism,* the Soul-theory, the belief that there is an individual mind in each living animal body; that between the mind and its organism a vital relationship holds; that the life processes are both mental and physical; that the directing force in evolution is to be found in the minds of the individual organisms, the urge of feeling in the lower, the increasing strength of emotion and will, with the widening scope of interest and of thought, in the higher organisms. Many arguments can be brought forward both for and against this theory, but we cannot discuss these here.

There has also been much discussion of what is called *The Two-Aspect Theory,* to which biological facts incline many inquirers. The theory assumes a psycho-physical being—a reality which we know under two aspects;

we think of the organism as one; as, while it lives, an indissoluble psycho-physical being. . . . The living creature gives an account of itself in two ways. It can know itself as something extended and intricately built up, burning away, moving, throbbing; it can also know itself as the seat of sensations, perceptions, feelings, wishes, thoughts. But there is not one process, thinking, and another process, cerebral metabolism (vital processes in nerve-cells); there is a psycho-physical life—a reality which we know under two aspects. Cerebral control and mental activity are, on this view, different aspects of one natural occurrence. What we have to do with is the unified life of a psycho-physical being, a body-mind or mind-body. The advantages of the two-aspect theory, if it is tenable, are that it does justice to the extraordinary intimate interdependence of what we may call "mental processes" and "brain-processes." It regards them as two equally real aspects of the continuous life of the organisms. . . . The objective side is the body as a living whole; the subjective side, in Man's case, is the unity of mind.[1]

In these days the now old-fashioned materialism of the previous generation, as Mr. Bertrand Russell says, "receives no support from modern physical science if, as seems to be the case, physics does not assume the existence of matter." We saw in a previous chapter ("The Foundation of the Universe") what the new view of the constitution of matter is. The atom of every element of matter is revealed as a particle of electricity; what electricity itself is we do not know. But we see how it comes about that the physicist tends to think of "matter" as less and less material. So does the chemist, and so the biologist. In that sense the old-fashioned "materialism" has gone.

The view of Mr. William James and others is that the "stuff" of the world is neither mental nor material, but, for the lack of a better name, a "neutral stuff," out of which both are constructed. Mr. Bertrand Russell, in his work *The Analysis of Mind*, endea-

[1] J. Arthur Thomson, *The System of Animated Nature.*

vours to develop this view as regards mental phenomena. We cannot sum up the problem better than another writer who says:

> Supposing we were able to understand all the phenomena—chemical, physical, physiological—of this intricate mechanism, we would be no nearer a solution of the problem of the connection between the objective and subjective impacts of the phenomena. . . . A philosophy which recognises both sets of phenomena—mental and physical—mutually adjusted and ever interacting, recognises the facts of the case, and does not delude the mind by offering a solution which is in reality no solution at all. The difficulty is somewhat lessened if we assume that behind all physical and mental phenomena there is a metaphysical essence, conscious or unconscious, and that the phenomena we term physical and mental are only different sides of the same kind. Such an essence can never be known to science, and the discussion of the possibility of its existence and of its properties belongs to the province of philosophy.

§ 3

Mental Processes

Psychology is the science of the mind, or more strictly, let us say, it is the science of the behaviour of living things; it includes the study of consciousness.

In the sense that the brain receives all those nervous impulses that result in consciousness, it would be true to say that the brain is the seat of consciousness. But that does not provide a solution of the problem of the origin of consciousness.

> No one doubts that consciousness has a material substratum, but the problem of the relation between the mental state and the molecular movements on nervous matter is as far from solution as in the days when little was known of the physiology of the nervous system. The old-fashioned method was to assign to the mind certain so-called faculties—perception, conception, imagination, reason, will—to explain the operations which they denote. The mind has not its will here, its

conscience there, and its reason somewhere else; it reasons, wills, and is conscientious as a whole. Thought, feeling, and will do not lie side by side, as it were, like stones in a mosaic, any of which could be removed without destroying the rest; they rather resemble the functions of the body, none of which are possible without the co-operation of all the others.[1]

Another way to describe mental activity was to regard every idea

as capable of existing in two conditions, or forms; on the one hand, it might be a conscious idea, or exist in consciousness; consciousness being spoken of as an illuminated chamber into which ideas enter in turn, to be lit up and active for a short period; and on the other hand, it might exist as an unconscious idea in the memory, a sort of Hades or dim underworld to which each idea, or its ghost, returns after its brief exposure to the light of consciousness; there to await and to seize any opportunity of emerging again into light and life. Within this underworld ideas remain linked together in complex groupings. The whole assembly of ideas, thus linked in the obscurity of memory, constitutes the structure of the mind; and mental activity consists in each idea dragging up after it into the light whatever ideas are linked or associated with it.[2]

When we come to the mind proper we may, using a purely pictorial analogy, regard it as consisting of three layers. The top layer we may call the region of the conscious life. It is, as it were, a vividly illuminated region, where everything that goes on is clearly seen. It is to this region that we normally refer when we seek the explanation of our conduct, and, as we shall see, the explanations we obtain in that way are often wrong. A little below this clear region is a semi-conscious region, a region which can become accessible to us by effort. It is in this region, for instance, that the information which is not present to our minds,

[1] F. S. Granger, *Psychology.*
[2] Wm. McDougall, *Psychology.*

but which we *can* remember, may be considered to be stored. Sometimes the contents of this region can be exhumed only by considerable effort, sometimes a very slight stimulus is sufficient. Beneath this layer, again, lies the region of the unconscious, and this region is, normally, quite inaccessible to our conscious mind. The description we have given is, of course, figurative, since we cannot suppose that the mind occupies space. But this division into layers is helpful in enabling us to understand the modern theories of the mind. The unconscious is the seat of the mental elements associated with the great primary instincts, and it is the great source of psychic energy. Of the activities going on in it we have no direct knowledge; we can infer something, however, as we shall see later, from observation, and more especially, according to some authorities, from *dreams*. The unconscious is the very basis of the psychic life of the individual.

The Importance of Complexes

Mental phenomena never occur singly, but always in some complex combination or another. It will help us in understanding the nature of the mind to consider it as a *network of mental elements*. Every mental element, every idea as we say, which comes into the conscious mind, calls up others. There are *associations of ideas,* to use the language of the older psychologists. It is because ideas are associated that we are able to go about our daily lives. If no ideas suggested any others, or if others were suggested purely at haphazard, we should never be able even to cross the road. A number of mental elements associated together so as to form some more or less loosely knit system is called a complex. To some men, for instance, the sight or sound of a typewriter may always, or usually, suggest to them an office; the smell of a certain flower may always bring back some early experience, and so on. Associations of this kind—associations of ideas, as it were—are called *complexes*. We may think, if we like, of ideas forming groups, and the whole of the contents of the mind as made up of groups of ideas—*complexes.* Further, complexes vary enormously in the emotional energy associated with them. Besides the great number of minor complexes brought about by a man's education, the nature of his work, and so on, there are so-called universal complexes. These are the complexes which centre round the three great primary instincts, or groups of instincts, and they are known as the sex-complex, the ego-complex, and the herd-complex.

Complexes which directly centre round a great primary instinct such as sex are associated with a great fund of emotional energy. The actual mental elements present in the sex-complex of any particular man, besides depending on inherited characteristics, depend also on his personal history. The ego-complex, associated with the primary instincts of nutrition and self-preservation, has most of its elements beneath the conscious level; and the same may be said of the herd-complex, which depends upon the gregarious instinct in man, and which plays an enormously important part in his life, as we shall see.

Amongst the three great universal complexes the ego-complex is the oldest and most profound. This is the complex with which is associated man's recognition of "his self." This very powerful complex may give rise to all sorts of unpleasant manifestations, to various exhibitions of greed and of the desire of self-aggrandisement; but it also gives rise to some of the most beneficent of man's activities. Amongst these we may mention the desire for *construction,* for the making of something which is a personal achievement, whether it be a house, a poem, or a system of philosophy. The desire to construct has certainly been one of the most potent factors in human advancement.

The Herd-Complex

The second great universal complex is the herd-complex, and this, as we have already said, depends upon the fact that man is gregarious. We do not know at what point in man's development he first developed the gregarious instinct. It must have been quite early, however, that man began to live in association with his fellows. The advantages bestowed by gregariousness are obvious. But the instinct of gregariousness brings with it certain consequences which are of the utmost importance in the psychic life of man. This instinct brings with it great *suggestibility*. The individual, as a member of the herd, must be very suggestible to impulses coming from the herd, in order to act in harmony with it. He must be able to yield unquestioning obedience to the voice of the herd. In the case of man his rational faculty, combined with his suggestibility as a gregarious animal, leads to the most diversified manifestations. The great bulk of man's opinions are in reality strictly non-rational, and are products purely of herd-suggestion; but that does not prevent him rationalising them. Many of them he does not trouble to rationalise. They appear to him "obivous"—as obvious as that good food is desirable; they come with *instinctive* force. The moral code in force in a community furnishes a set of beliefs of this kind. This set of beliefs changes from time to time and from country to country, but whatever set of beliefs may be in vogue in any particular community at any particular time is "obviously" right.

Two Main Types

We cannot consider in detail the manifestations of the three great groups of primary instincts, but we may discuss, for a moment, two types, in one or other of which nearly every human being can be classed. These two types of human beings are called by Mr. Trotter the *stable* and *unstable* types.

The *stable* type is the type which is often described as forming the backbone of the country. A man of this kind is energetic, strong-willed, and full of settled convictions. He is perfectly at home with the laws and traditions of the community of which he is a member. His aims are of the kind that the community as a whole can understand and approve, and he is steadfast in his pursuit of them. He has decided views on moral questions, and on political and any other subjects. He is never in doubt as to what is right and what is wrong.

The great drawback to this type is its insensitiveness to experience: it is incapable of surveying any question from an entirely fresh standpoint. Indeed, it is apt to regard the searching questioning of accepted and established things, such as a code of moralities or a system of politics, as either foolish or wicked or both. Great changes in current practice and ideas, however desirable such changes may be, cannot be effected by the class of people—and it predominates in numbers—which has the strong prevailing gregarious instinct—in other words, in which the herd-complex is strongly engrained.

The *unstable* type has qualities almost exactly opposite to those of the stable type. Thus, a man of this type has very few settled convictions, although he may have plenty of enthusiasms. He can easily be won to a new cause, and he as easily falls away therefrom. He may undertake a number of projects, but it is unlikely that he will persist with any one of them long enough to carry it to a successful conclusion. He has what is called a weak will, and he can by no means accept the ruling of the community on all questions. His great positive merit is his sensitiveness to experience, and, indeed, it is from this that all his trouble springs. He is always changing his mind because he is always open to fresh impressions. He is, usually, the intellectual superior of the stable type, although the stable type often despises him. But each type has its great disadvantage, and neither represents what a human being could and should be.

Conflicts

The fact that different complexes may be incompatible with one another leads us to the important question of *conflict*. A perfectly healthy mind is a mind which has established complete harmony between its different complexes. But the perfectly

SIR EDWIN LANDSEER'S PAINTING "A NAUGHTY CHILD"

A fine study that will suggest much to the student of psychology.

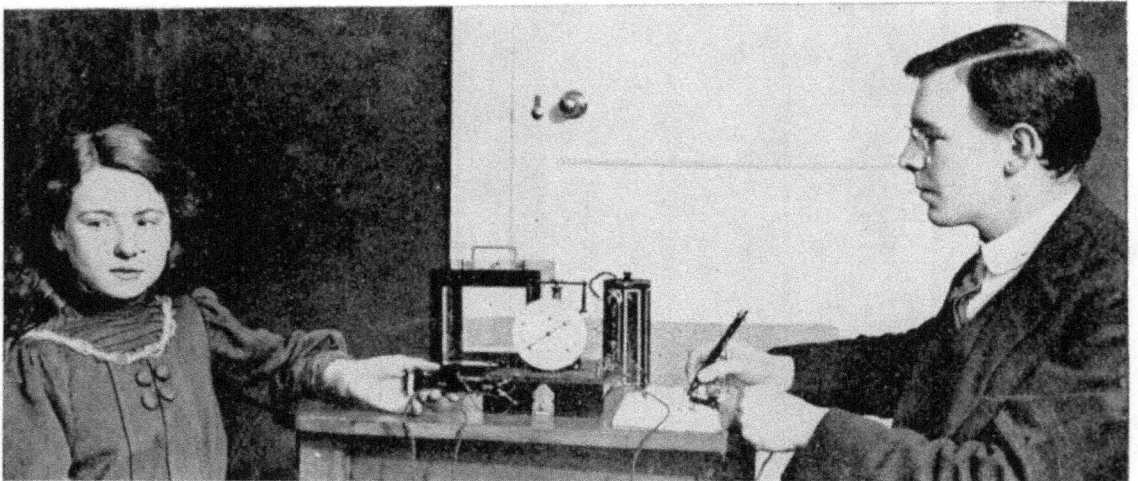

MR. CYRIL BURT, PSYCHOLOGIST TO THE LONDON COUNTY COUNCIL, MEASURING THE SPEED OF THE THOUGHT OF A
CHILD WITH A CHRONOSCOPE TO TWO-HUNDREDTHS OF A SECOND

healthy mind, in this sense, is very rare; we usually find that several of a man's complexes are incompatible with one another, and on those occasions when more than one are aroused there is *conflict* between them. Thus it may often happen that a man's "selfish" desires, those springing perhaps from his ego-complex or his sex-complex, conflict with the moral code of the community, a code which has great weight with him because it is associated with his herd-complex. Such conflicts are favourite themes for novelists: the father torn between patriotism and his love for his son; the intending monk torn between his religion and his love of his family; the man torn between an illicit love passion and his sense of morality. Conflict plays a prominent part in the psychic life of most people, and it leads to very important consequences. For the conflict must be settled, and there are two very important ways of settling it. There is the method of *rationalisation*. One of the conflicting complexes is allowed to triumph, but not consciously. Reasons are invented for the resultant action which have nothing to do with its psychic causes, but which prevent the man from feeling "ashamed" as we say. Thus a primitive brutal desire for revenge may be disguised as justice. An exhibition of ruthless greed—as in some unscrupulous business deal, for instance—will be explained by pointing out that it is for the good of the community that its most efficient citizens should come to the top, and so with other conflicts.

Another very important method of settling a painful conflict is by *repression* of one of its factors. It is this method which has been chiefly studied by Freud, and he has succeeded in showing how very great its importance is. A man decides completely to ignore one of the conflicting complexes—he puts it out of his mind. But, as Freud has shown, the ignored complex is not thereby destroyed. It is repressed into the unconscious, but it is still energetic and may manifest its existence in a number of ways, ranging from certain phenomena of *forgetfulness* down to hysteria and insanity. It may happen, for instance, that the repressed complex leads to a certain kind of forgetfulness, a forgetfulness of those things with which it is associated. A man may forget an appointment from which he anticipated something unpleasant, he may forget the existence of unpaid bills. Such cases are cases of *active* forgetting, and are to be distinguished from cases of *passive* forgetting, where the matter is forgotten simply because it made very little impression on the mind. A slip in speaking or writing may sometimes testify to a repressed complex; the substituted word corresponding to a wish, but a repressed wish, of the speaker or writer, as when the President of the Austrian Lower House announced that the sitting was closed when he should have said it was opened, the reason being that he privately expected no good from the sitting and would have liked it closed.

§ 4

PSYCHO-ANALYSIS

Professor Freud's Theories

A comparatively new branch of psychology is that closely associated with the work of Professor Freud of Vienna. It deals mainly with the phenomena of the unconscious. Whatever may be said of Freudian theories, they have at least opened up a wide field of study. Part of Freud's doctrine has become fairly well established; on the other hand, a great deal of it is regarded as merely ingenious theory, which is not generally accepted. This "new" psychology is of very great interest, because of the bearing it has on medical practice and the work of the teacher.

The chief theory of the Freudian psychology is, that there is a great part of the mind of which we are unconscious; that this unconscious part exercises an enormous influence upon our thoughts and actions, without ourselves being aware of it. Freud conceived the idea that the influence of the unconscious mind was especially active as a cause of dreams, and thus he was led to his now familiar theory of the interpretation of dreams.

The work of Professor Freud, his disciples and his critics, has thrown a flood of light upon the working of the human mind, and led to curious alterations of our views upon dreams, insanity, myths, art, and religion. In dealing with patients who were suffering mainly from functional diseases of the nervous system, Freud found that what had been regarded as the symptoms of the disease, such as paralysis of the limbs, blindness, deafness, and mutism, were frequently connected in some definite way with the original onset of the disease; blindness, for example, might date from some violently painful occurrence of which the patient had been a witness. This connection was not as a rule recognised by the patient's waking consciousness, but it revealed itself occasionally to the doctor when the patient was hypnotised; sometimes also it was brought out by the dreams which the patient described; but in general the ordinary consciousness of the subject resisted all attempts to probe back to the original cause of the disease.

Turning his attention to dreams, Freud found that in the case of normal individuals also there were painful experiences, never revived in the fully conscious mind, but playing a great part in the dreams of the subject, appearing there in a more or less disguised form; and that the interpretation of the dream in both normal and abnormal subjects invariably led back to some wish or desire of the individual, which it was impossible for him, for physical, moral, or social reasons, to realise in waking life. The dream was the mimic realisation of the wish.

The instinctive or voluntary forgetting Freud called Repression; the repressed ideas were not, however, destroyed, but were constantly endeavouring to force their way back into consciousness. He gave the name of the "unconscious" to the mass of repressed memories of all kinds. For the repression of a wish involves also the repression of the whole system of experience to which the wish belongs; hence, for example, the fact that we can rarely remember our infancy-time at all.

The Subconscious

We have all some experience of what is called *subconsciousness;* an idea, as it passes to and from the focus of consciousness, gradually becomes clear and vivid, then fades away into dimness and vagueness, till it is merged in the general mass of feeling and loses all distinctiveness; a word is "on the tip of the tongue," later it is clearly thought and spoken. I have an appointment to remember. I do not think of it for hours, and then—in good time, perhaps—it walks into my "consciousness." I resolve to awake at six in the morning, and—if my mind is of the right kind—as the clock strikes six, or just before it, I awake. These are different cases in which an idea, a thought, is apparently not in consciousness, and yet not wholly out of it. The term "subconsciousness" has been used for this class of phenomena, where, apart from the "dominant" or "personal" consciousness, certain strands of experience, which have once been conscious, continue somehow to live, and in due time make their influence felt in the dominant consciousness.

The theory of Freud is, that in the unconscious part of the mind there lie dormant memories of the past and especially "repressed" impulses. These repressions represent the resistance we make to a wish or impulse which we think we ought not to satisfy, bcause it conflicts with some other interest; or they mean the effort we make to put out of our mind some unpleasant memory. The effort to repress may not be deliberate, it may be unconscious repression. In any case there may be a repression to such an extent that the memories pass entirely from us, or as it is held, they are pushed deep into the unconscious, where they continue to exist. We are asked to believe that "the unconscious includes many impulses and memories which remain buried in the depths of the mind," and that they persist in trying to return to the living mind. Further, it is said that to some extent they do so, influencing the mental life even although we are not conscious of the influence at work. In this way repressed tendencies are supposed to get a partial satisfaction.

§ 5

Cases of Mental Disorders

The records of medical men in their work connected with nerve cases in military hospitals during the war has provided much material for the study of abnormal psychology of this kind. Cures of paralysis of various organs, of morbid obsessions, and unreasonable fears have been recorded and described by responsible members of the medical profession. The origin of many mental troubles has been traced to repression of disturbing emotional experiences, bygone and forgotten by the patient. The recalling or revival of such lost memories of patients by medical men skilled in psychopathology have, by clearing the mind of the patient, enabled physicians to effect many striking cures of mental disorder.

The theory is that the bringing to light and the re-living of the suppressed emotional experiences is a means of getting rid of excessive emotion. The patient is enabled to assume a new attitude towards them. By way of illustration we may give one instance:

The following case of the influence of forgotten experience is described by Dr. W. H. Rivers in the *Lancet,* and we take this excellent summary of it as given by Professor Valentine in his *Dreams and the Unconscious.*

It is the case of a young medical officer, who even before the war had a horror of closed-in spaces, such as tunnels and narrow cells. He would never travel by the tube railway, and was seized with fear in a train which passed through a tunnel. One can imagine his intense distress when on entering a dug-out he was given a spade and told it was for use in case he was buried alive. His sleep was greatly disturbed, and his health became so bad that he was invalided home. Instructions to keep his thoughts from the war and to dwell exclusively on pleasant topics proved useless. He had terrifying dreams of warfare, from which he would awake, sweating profusely, and thinking he was dying. At this stage he came under the care of Dr. Rivers. The patient was asked to try and remember any dreams he might have and to record any memories which came into his mind while thinking over the dreams. Shortly afterwards he had a dream, and as he lay in bed thinking it over there came into his mind an incident which seemed to have happened when he was about three years of age, and which had so greatly affected him at the time that it now seemed to the patient almost impossible that it could ever have been forgotten. He recalled that, as a little boy, he and his friends used to visit an old man in a house near his own, and to take him odd articles discarded at home, in return for which they received a copper or two. On one occasion he went alone, down the long, dark passage leading to the old man's home, and on turning back found that the door at the opening of the passage had banged to, and he was unable to escape. Just then a dog in the passage began to bark savagely, and the little child was terrified, and continued so until he was released. After another dream the patient woke up to find himself repeating "McCann! McCann!" It occurred to him, suddenly, that this was the name of the old man. Inquiry of the parents of the patient revealed the fact that an old rag-and-bone man had lived in such a house as the patient remembered, and that his name was McCann.

The result of this recovery of memory, with the explanation of his abnormal fears of closed-in spaces, had a great effect on the patient. A few days afterwards he lost his fear of closed-in spaces, and he afterwards travelled in tube railways and tunnels without discomfort. Indeed, he was so confident of himself at once that he wished Dr. Rivers to lock him up in some subterranean chamber of the hospital as a proof of his cure. The particular point to be noticed here is that an entirely forgotten experience continued, apparently, to have an influence upon conscious mental life. Other points of interest are these: that the original experience was an intensely emotional and disturbing one; that the experience was recalled through reflecting on a dream; that the conscious effort of will to banish the unreasoning fears had no effect; that the fearsome experience, though repressed until forgotten, found its way out to consciousness through the repeated emotions of fear. This constant fear was stimulated by being in closed-in spaces, that is, by situations similar to the original one, though that was forgotten.

There are many such cases as this on record. A great deal of work has been done on similar lines, and the study of disorders of various kinds, having a mental origin, has been put on a scientific basis within the last few years. This is not the place to describe the methods of the practitioner; the principles followed depend on individual cases.

§ 6

Dreams

Much, probably far too much, has been made of the claim that psycho-analysis may be applied to the interpretation of dreams. The starting-point from which Freud's theory was developed was the interpretation of dreams, based on the assumption that dreams are the symbolical expression of repressed tendencies. To claim that every dream is determined by the subconscious working of a repressed tendency is unwarrantable, and the theory is not accepted by those most qualified to speak on the subject. On the other hand, it would be an extreme view, as Dr. William Brown says, to deny all meaning to dreams, and regard them as merely the confused and jumbled reappearance during sleep of memories belonging to the person's past history, strung together in any chance order.

The recent work on dream analysis, however, has added immensely to our knowledge, and we now possess a theory which undoubtedly covers a very large part of dream phenomena, even although it certainly does not cover the whole. This theory is, briefly, that a dream is the symbolic fulfilment of a repressed wish; the wish has been repressed because, for one reason or another, its appearance in the conscious mind is attended with pain. But, as we have seen, repressed elements do not lose their vitality; they continue to work and they endeavour, as it were, to manifest themselves in some way or another. Now during sleep the barriers between the conscious and the unconscious are to some extent relaxed. Elements which are ruthlessly repressed in the waking life are now subjected to a less severe repression. But these elements cannot emerge in their naked purity, as it were; they exhibit themselves in a disguised form, often of the most fantastic description. In this way the wish secures a partial satisfaction. In his book on *The Interpretation of Dreams,* Freud gives a large number of such cases of symbolic fulfilment, and explains the technical processes by which these dreams are related to forgotten episodes in the life of the patient. Many of these cases are more ingenious than convincing.

Not all dreams are due to repressed wishes. Many dreams are more or less inchoate reproductions of impressions received during the day; such dreams, however, have a fragmentary character. In very many cases where the dream is a rounded and completed whole it is also an allegory, a symbolic manifestation of elements which have been repressed into the unconscious. The repressed elements, even so, do not secure complete fulfilment. Repression is still operative, although it is relaxed. There is still what Freud calls the "censor." Dreams may illustrate very interestingly, in fact, the indirect ways in which psychic energy seeks an outlet, when direct satisfaction is for some reason or other denied it. Many works of art are similar to dreams in this respect. In some cases of very deep and powerful repressed complexes, a dream fulfilment may not be satisfactory. An actual pathological condition may be set up; hysteria, insanity, and dissociations of personality, as in certain well-known cases of double personality, may be caused by the repressed complex. Many cases of this kind were brought into being by the terrible psychic strains of the war.

It is admitted that a certain class of dreams may be possible of interpretation, but we cannot discuss the subject further here; it cannot be accepted that Freud's theory of repression accounts satisfactorily for all dreams.

Another view is that which regards dreams in quite a different light. Dr. William Brown puts it in these words:

> The function of a dream is to guard sleep. Sleep is an instinct like fear, flight, and the rest, and has a function which has developed in the course of evolution. At night this instinct of sleep comes into play, but it finds itself in conflict with other instincts and tendencies, as well as with external impulses. Desires, cravings, anxieties, the memories of earlier days, all of which are the lower and fundamental elements of the mind, well up and strive towards consciousness, while the main personality is in abeyance. If they reach consciousness sleep is at an end, but the dream, which is a sort of intermediary form of consciousness, intervenes, and makes the impulses innocuous, so that sleep persists. This theory covers the entire ground of all types of dreams.

There are other aspects of abnormal psychology which imply subconscious operations with which we have not dealt. The subject of telepathy, clairvoyance, materialisations, and other phenomena which appertain to psychic experience will be discussed by Sir Oliver Lodge in the following chapter.

BIBLIOGRAPHY

FREUD, *Interpretation of Dreams.*
GREEN, *Psychanalysis in the Class Room.*
LLOYD MORGAN, *Comparative Psychology.*
Low, *Psycho-Analysis.*

McDougall, *Psychology, Social Psychology,* and *The Group Mind.*
Myers, *Experimental Psychology.*
Tansley, *The New Psychology.*
Titchener, *Text Book of Psychology.*
Trotter, *The Herd Instinct.*